Hymnwriters 2

Bernard Braley

STAINER & BELL - LONDON

First published in Great Britain by
Stainer & Bell Ltd, 82 High Road, London N2 9PW

British Library Cataloguing in Publication Data

Braley, Bernard 1924 –
Hymnwriters
2
1. Christian Church. Public worship. Hymns. Words
English
I. Title
264'.2'0922

ISBN 0 – 85249 – 763 – 6
ISBN 0 – 85249 – 764 – 4 Pbk

Printed in Great Britain at the Alden Press, Oxford

Contents

Preface	v
List of Illustrations	ix
John Newton	1
James Montgomery	59
Henry Baker	121
Albert Bayly	147
Index of Persons	182
Index of Place Names	187
General Index	191
Index of Tunes	202

Preface

Like *Hymnwriters 1* this book is first of all just to be enjoyed for the reading of it. Although basically concerned with the literal texts by the chosen authors, many tunes that have been sung to their words have been mentioned and these provide for the opportunity of an occasion performing a writer's hymns as his story is told. Once again, the imaginative will see chances for dramatisation and costume in presenting these stories; and those preparing sermons and liturgies will find many useful incidents, quotations and stimulating ideas relevant to our present age.

John Newton, slave-captain turned Anglican parson, wrote prolifically; and Moravian poet and journalist produced even more paper. The crowding of quotations here and there portray accurately the weight of words with which they surrounded their contemporaries from time to time. Sir Henry Baker's story is at the same time the beginning of the story of *Hymns Ancient and Modern*, in use in its several editions in many Anglican churches across the world for more than a century. The chapter on the twentieth century Albert Bayly, essentially an ecumenist but ordained as a Congregationalist minister who rejoiced in the merger which formed the United Reformed Church in the United Kingdom, contains less quoted material from his poetry and hymns than the other chapters. This arises from considerations relating to copyright and because his poems and hymns are being published in a complete edition about the same time as this book (David Dale: *Rejoice: the Hymns and Poems of Albert Bayly;* Oxford University Press) and readers wishing to explore his corpus further are referred to this collection.

At the request of several correspondents communicating about *Hymnwriters 1*, I have extended the General Index to include all quoted texts by the four hymnwriters whose lives are traced in this volume.

I am obviously indebted to earlier biographers in preparing the first three chapters. For the chapter on Albert Bayly, I am especially indebted to his widow, Mrs Grace Bayly for valued information and insights and permission to quote from her husband's notes on his dream of the English Church in the 21st century and from a letter he wrote to the local press in Whitley Bay. Notes on Albert Bayly's life and work prepared by Mr Edmund Jones were also most helpful.

I am grateful to Mrs Jean Lowe and Mr Keith Wakefield for their extensive work in layout and design. Mr Wakefield's expertise in the field of photography has been especially valued. Every effort has been made to trace copyright owners who are acknowledged below, but in a few instances of old letters and articles, this has not proved possible despite considerable enquiry.

Thanks are especially due to the staff of the British Library, of Haringey Public Libraries, of Sheffield City Libraries, of the Congregational Library, of the Northumberland Record Office, the Tyne and Wear Archives Service and the Herefordshire and Worcestershire Records Office. Special thanks are due to Miss Sylvia Bull, the Curator, and to the Trustees of the Newton and Cowper Museum, Olney, for access to many papers, and to the Vicar of Olney for permission to photograph stained glass windows in the Church. Readers are warmly commended to visit the church and museum at Olney if they can.

I am indebted to the Proprietors of *Hymns Ancient and Modern* and especially Mr Gordon A. Knights for access to archives and in particular for permission to print extracts from the Proprietors' Minute Books, notes of a meeting with the Dean of Canterbury, and transcripts of or extracts from the following letters:

From Arthur Sullivan to Sir Henry Baker concerning use of Sullivan's tunes
From Reverend William Walsham How to Sir Henry Baker on the subject of copyright
From Reverend John Keble to the compilers of 'Hymns Ancient and Modern' giving his advice
From Sir Henry Baker on 15th December 1858
From W. H. Monk concerning Sir Henry Baker's interest in 'The Psalter'
From John Stainer to Sir Henry Baker concerning expression marks
From John Stainer to Sir Henry Baker sending proofs
From W. H. Monk to Sir Henry Baker concerning Monk's remuneration
From Caroline Noel to Sir Henry Baker concerning 'At the Name of Jesus'
From Reverend John Padfield to the 'Proprietors of Hymns Ancient and Modern' concerning the music of Monkland Church

Thanks are due to Oxford University Press for permission to quote hymns and poems written by Albert Bayly, including stanzas by Brian Wren of *Lord of the boundless curves of space*, the material being printed on Pages 146, 148, 153, 163 to 167, 169, 172 to 181; to Reverend Dr Fred Pratt Green for permission to print his *In Memory of Albert Bayly*; to Mansfield College, Oxford for raising no objection to the printing of extracts from the *Mansfield College Magazine* and to Mr John Creasey and The Congregational Library for access to their set of this publication; to the *Burnley Express and Burnley News* for permission to reprint extracts from articles by Albert Bayly written for their paper; to Trinity United Reformed Church, Whitley Bay for permission to quote from a Minute Book of Park Avenue Congregational Church and to Mr Bruce McIntosh, Secretary of the Church for useful data; to *Morpeth Herald* to quote from their columns reporting events involving Albert Bayly; to the Northumberland District Council of the United Reformed Church in the United Kingdom for permission to reprint an extract from *One Hundred Years of Congregationalism in Morpeth* and to quote from the records for 1942 of Dacre Street Congregational Church, Morpeth; to the elders of Thaxted United Reformed Church for permission to quote from the newsletters of Thaxted Congregational Church during Albert Bayly's ministry and to Miss Ruth Bennett for information and access to papers; to Miss Ruth Bennett and Mrs Joy Duport nee Ashford for permission to print their copyright hymn tune *Broomfield*; to Eccleston United Reformed Church, Saint Helens for permission to quote from Newsletters of Eccleston Congregational Church and to Miss H. M. Gabbott for her research on my behalf in those newsletters and the Minute Books of the Church; to the Council for World Mission as successors of the London Missionary Society for permission to quote from Albert Bayly's book *Kendall Gale* and to print an extract from the pageant *God's Building*; to the Parish Council of Monkland Parish Church for permission to print short extracts from the Minutes of Vestry meetings; to the late Prebendary John Clingo for helpful information about Monkland Church, Vicarage and School; to Mr and Mrs Morgan for access to the Old Vicarage; to the Epworth Press for permission to quote from *The Journal of a Slave Trader*; to Dr C. C. G. Rawll and the Reverend Peter Galloway for sight of a part of their manuscript *Lives and Vicars of All Saints*; to the Honorary Librarian of the United Reformed Church History Society, to

Reverend R. Parker of the Lancashire Congregational Union and to Mr R. Helme, the Church Secretary of Bethesda Street Congregational Church for tracking down information; and for stories, not needing copyright clearance but providing valuable information, in the columns of the *Hereford Journal* and the *Whitley Bay Seaside Chronicle and Visitors' Gazette.*

The publishers would like to thank the following for their kind permission to reproduce the photographs in this book:

Mrs Grace Bayly: 147, 177
British Library: 58, 61
Lancashire Library Burnley District: 165
National Portrait Gallery, London: 23
Northumberland District Council of the United Reformed Church: 161
Sheffield Public Libraries and Record Office (Archive SLPS222): 102
Keith Wakefield: xii, 21, 41, 57, 89, 119, 120, 124, 126, 128, 134, 159, 163, 168, 170

Bernard Braley
July 1989

List of Illustrations

Page xii Stained glass window (John Newton), Olney Parish Church

Page 1 Ships under construction and repair, Deptford c.1730

Page 9 John Newton

Page 13 Plan showing how slaves could be packed in a ship for maximum profit

Page 19 John Wesley

Page 21 Stained glass window in Olney Parish Church

Page 23 George Whitefield

Page 41 St. Mary Woolnoth, City of London

Page 49 Score of 'Comfort Ye' from 'The Messiah' Handel

Page 51 Interior of St. Mary Woolnoth – prior to alteration

Page 53 St. Mary Woolnoth

Page 57 Bank Station booking hall – former site of crypt

Page 58 Montgomery's birthplace at Irvine

Page 61 The house in which Montgomery lived at Wath upon Dearne

Page 79 James Montgomery

Page 89 Inscription at base of Montgomery Memorial, Sheffield

Page 98 Isaac Watts

Page 99 B. Huntsman & Co, Sheffield – Steel Furnaces and Offices, c.1830

Page 102 James Montgomery and the Gale sisters in silhouette

Page 119 The Montgomery Memorial, Sheffield

Page 120 Memorial lich-gate, Monkland Parish Church

Page 122 Henry Baker

Page 124 The former Vicarage, Monkland

Page 126 Monkland School – now the Village Hall

Page 128 The Clergy House, St. Barnabus, Pimlico

Page 134 Monkland Parish Church

Page 138 Musical extract in the hand of John Stainer

Page 146 Music of BROOMFIELD

Page 147 Albert Bayly

Page 151 Whiteley Bay Co-operative Society Limited, c.1920

Page 153 Albert Schweitzer

Page 156 Bridge Street, Morpeth, c.1930

Page 159 Dacre Street Church, Morpeth

Page 161 Church Training Corps minutes – January 1944

Page 163 Lichfield Cathedral

Page 165 Burnley in the 1950s

Page 168 Swanland Congregational Church, Humberside

Page 170 Eccleston United Reformed Church, St. Helens

Page 177 Thaxted, drawn by Albert Bayly

Page 181 Tune THAXTED by Gustav Holst

Stained glass window, Olney Parish Church

John Newton

John Newton was the son of diverse temperaments. His father was a ship's master, engaged in the Mediterranean trade, whose upbringing included childhood education in Spain. This schoolday experience taught him to take a pride in his appearance and he was probably the best-dressed sea captain in port. By contrast, John's mother was a practising religious dissenter, whose nature it was to be quietly reverent and humbly obedient to the Lord.

John's birth year of 1725 co-incided with one of the quieter periods of Western European history: it was a good time to be a maritime trader. In his early years, as his father was usually away at sea, John's upbringing was under his mother's influence. He was introduced early to the practice of prayer and to the Bible stories: John's mother was a skilled teacher and her son a quick learner, reading widely at the age of four.

A few days before John's seventh birthday, his mother died and he was left in the care of the servants while his father was on his travels. His father soon married again, this time to the daughter of a prosperous glazier, so John was packed off to a boarding school in Essex. Apart from cruel bullying, endemic at the time to such establishments, Newton recorded in his autobiography that the imprudent severity of the master almost broke his spirit and undid the learning he had received from his mother. After two years, a new usher succeeded in giving him a little Latin: but finding the lad had a rapid brain, pushed him too fast and failed to give him a good grounding in the language.

My mother (as I have heard from many) was a pious experienced Christian . . . a Dissenter . . . I was her only child; and as she was of a weak constitution, and a retired temper, almost her whole employment was the care of my education. I have some faint remembrance of her care and instructions. At a time when I could not be more than three years of age, she herself taught me English; and with so much success . . . that when I was four years old, I could read with propriety in any common book that offered. She stored my memory, which was then very retentive, with many valuable pieces, chapters, and portions of Scripture, catechisms, hymns and poems.

John Newton: From 'An authentic Narrative of some remarkable and interesting Particulars in the Life of John Newton, communicated in a Series of Letters to the Rev. T. Haweis: Occurrences in early Life'

Ships under construction and repair, Deptford, c. 1730

Whilst he went home to his step-mother in the holidays, he felt unwanted and was content enough to end his school days at the age of ten. He gained his sea-legs, voyaging with his father several times in the next few years: whilst life for the master's son was comparatively comfortable, the language and outlook of the crew was dramatically disturbing for a lad brought up by a gentle mother, pious in the extreme. A great gulf existed between father and son; when John was fifteen, his father decided it was time for some outside discipline. An old contact in Alicante offered a job: and the adolescent Newton became the employee of a Spanish merchant. But John was a rebellious teenager who did not take kindly to obeying his boss's instructions and before long, he was back on board with his father.

I was fond of reading from a child; among other books, Bennet's *Christian Oratory* often came in my way: and though I understood but little of it, the course of life therein recommended appeared very desireable; and I was inclined to attempt it. I began to pray, to read the Scripture, and keep a sort of diary. I was presently religious in my own eyes; but, alas! this seeming goodness has no solid foundation, but passed away like a morning-cloud, or the early dew.

John Newton writing of childhood religion in the letter revealing 'Occurrences in early Life'

In all these years, John Newton had a love-hate relationship with religion. This was heightened at one point by involvement in a riding accident: John wondered how it would have been if he had lost his life and summoned to appear before God. He recalls in his autobiography taking up some kind of religious commitment several times only to backslide. He tried to understand writings he found in *The Christian Oratory*. The perusal of the *Family-Instructor* put him *upon a partial and transient reformation*. On the last occasion, following the practice of a strict sect, he lived like a pharasee, becoming as a result gloomy, stupid, unreasonable and useless. This introspective lad could not have been a very reliable seaman with his mind so much on philosophical matters! On one voyage, John picked up a volume by Lord Shaftesbury which questioned the religious dogma of the time. Whilst to many such questioning would have been blasphemous, John appreciated there were issues to be considered. Soon John knew *The Moralists* off by heart. The free-thinkers had added their influence to the pot-pourri of adolescent experience.

I would resolve for the future, by thy Grace, to exercise more Love and Benevolence towards Mankind to mortify my self, to put off that undue Self-love, that has been the Occasion of so much Injury to others, and Dishonour to my Profession. O! that I may love my Neighbour as myself.
I would resolve to set this Rule constantly before me in my Intercourse with Mankind; consider myself in them, and offer them nothing I should not like myself; or, according to the Laws of impartial Reason and Equity, I could object against; I would bind it about my Neck, write it on the Posts of my House, my Shop, or rather beg that God would write it on the Table of my Heart.

Benjamin Bennet: From 'The Christian Oratory: Of Meditation'

I shou'd think, said he (carrying on his Humour) that one might draw the Picture of this *Moral* Dame to as much advantage as that of her *Political* Sister; whom you admire, as describ'd to us 'in her *Amazon-Dress*, with a free manly Air becoming her; her Guards the *Laws*, with their written Tables, like Bucklers, surrounding her; Riches, Traffick, and Plenty, with the *Cornucopia* serving as her Attendants; and in her Train the *Arts* and *Sciences*, like Children playing' – The rest of the Piece is easy to imagine: 'Her triumph over Tyranny, and lawless Rule of Lust and Passion' – But what a Triumph wou'd her Sister's be! What Monsters of savage Passions wou'd there appear subdu'd! There fierce *Ambition, Lust, Uproar, Mis-rule*, with all the *Fiends* that rage in Human Breasts, wou'd be securely chain'd. And when *Fortune* her-self, the Queen of Flatterys, with that Prince of Terrors, *Death*, were at the Chariot-wheels, as Captives; how natural wou'd it be to see *Fortitude, Magnanimity, Justice, Honour*, and all that generous Band attend as the Companions of our inmate Lady *Liberty*! She, like some new born Goddess, wou'd grace her Mother's Chariot, and own her Birth from humble *Temperance*, that nursing Mother of the Virtues; who like the Parent of the Gods (old Reverend *Cybele*) wou'd properly appear drawn by rein'd Lions, patient of the Bit, and on her Head a Turret-like Attire: the Image of defensive Power, and Strength of Mind.

Anthony Ashley Cooper, Third Earl of Shaftesbury: From 'Characteristic Treatise V viz The Moralists, a Philosophical Rhapsody'

. . . it was a poor religion; it left me, in many respects, under the power of sin; and, so far as it prevailed, only tended to make me gloomy, stupid, unsociable, and useless.

From John Newton's writing of trying to live a strict religious life in the style of a Pharisee later in the letter revealing 'Occurrences in early Life'

By 1742, John's father, himself not planning to go to sea again, wondered how to settle this dreamer of a son into the real world. The retiring sea-captain consulted Joseph Manesty, a business friend in Liverpool, who suggested employment on a sugar plantation in Jamaica. A planter's pay, with *almost* honest extras, could provide enough for a young man to arrive home in due course with plenty of funds. John was unenthusiastic but, at least, it would take him away from father and stepmother.

Many years earlier, there had been two childhood friends. One was John's mother: the other was the mother of Mary Catlett. It had been an abiding relationship and indeed it was in this friend's home that his mother had died. Now, after many years of silence, a letter arrived from the Catletts. It invited John to call in to their home near Chatham if his travels ever took him in that direction. While waiting for the ship that was to transport young John to Jamaica, his father asked him to journey to Maidstone to transact some business.

On the way home, John took up the invitation and stayed for three weeks. He had fallen head over heels in love with the eldest daughter of the household, fourteen-year old Mary. Infatuation banished all thoughts of Jamaica. Eventually he returned home: and his disappointed father fixed up for him to serve as an ordinary seaman before the mast in a friend's ship bound for Venice: quickly such scruples as remained from his early pious upbringing were thrown overboard as he cussed and swore like the rest. The voyage over, he made quickly for Mary's home: but his present employment hardly allowed them to become engaged, though he was made a

No thing could be more suited to the romantic turn of my mind, than the address of this pompous declamation. Of the design and tendency I was not aware: I thought the author a most religious person, and that I had only to follow him and be happy. Thus, with fine words, and fair speeches, my simple heart was beguiled. This book was always in my hand: I read it till I could very nearly repeat the Rhapsody *verbatim* from begin-- ning to end. No immediate effect followed; but it operated like a slow poison, and prepared the way for all that followed.

John Newton writing of Lord Shaftesbury's 'Characteristics' later in the letter revealing 'Occurrences in early Life'

welcome visitor. Then the Navy intervened. While walking by Chatham Harbour, he was press-ganged into service on *HMS Harwich* bound for the East Indies.

He was a man of remarkable good sense, and great knowledge of the world: he took great care of my morals, but could not supply my mother's part . . . he always observed an air of distance and severity in his carriage, which overawed and discouraged my spirit. I was always in fear when before him, and therefore he had the less influence.

John Newton writing of his father in the letter revealing 'Occurrences in early Life'

He suffered at once the severe hardships of a pressed recruit: his father, believing that harsh discipline might be no bad thing for his son, made no immediate move, but in due time made a recommendation to the captain on the basis of young Newton's sea-going experience and the lad was taken on quarter-deck as a midshipman. Here he records in his diary that he met with companions who completed the ruins of his principles. When the ship neared Plymouth, he was sent, in charge of a group of men, on a mission to the beach with orders to ensure that none deserted. Knowing that his father was in the Torbay area, he chose to desert himself: his dislike of Navy life and the attraction of Mary Catlett sparked off this impetuous act which was soon thwarted by the military. He was arrested, returned to his ship, put in irons, flogged with the cat o' nine tails and demoted. Life with the lowest of the low was now worse than ever; for his new mates had been forced to take orders from Midshipman Newton and valued the opportunity to get their own back.

I walked through the streets guarded like a felon – My heart was full of indignation, shame, and fear. – I was confined two days in the guard-house, then sent on board my ship, kept a while in irons, then publicly stripped and whipped; after which I was degraded from my office, and all my former companions forbidden to show me the least favour, or even to speak to me.

John Newton writes of his capture and punishment after deserting from his ship in the fourth letter of 'An authentic Narrative' : 'Voyage to Madeira'

Naval captains sometimes chose to rid themselves of trouble-makers by swapping them for crew from a merchantman. When *HMS Harwich* was lying off Madeira, such an opportunity arose and John found himself aboard a ship engaged in a trade which many still thought respectable and proper. He had joined a slaver. Quickly his natural rebelliousness made him popular with the crew: but he went too far in writing a song which delighted the hands for the fun it poked at the captain. This man's sudden death brought a temporary reprieve from punishment: but he was alarmed that he would be handed back to the Navy at the first opportunity.

A passenger on the ship by the name of Chow had made his fortune selling poor quality goods to African chiefs in exchange for the slaves they rounded up. Chow thought Newton might make a useful employee and the mate who had taken over on the captain's death was not sorry to discharge him.

As the employee of a slave trader, perhaps there would be the chance to make enough money to return and marry Mary. The work Chow had in mind for Newton was to lead expeditions further inland when slaves ran short: and there to capture or buy more hapless men, women and children. Chow had a native woman as his common law wife and she

How little they thought it was he,
Whom they had ill treated and sold!
How great their confusion must be,
As soon as his name he had told!
I am Joseph, your brother, he said,
And still to my heart you are dear,
You sold me, and thought I was dead,
But God, for your sakes, sent me here.

John Newton: From 'Joseph made known to his Brethren' which begins 'When Joseph his brethren beheld'

was jealous of the way the newcomer took away her husband's attention. He was glad to have a white man for company and gave much less time in consequence to his wife. When Newton fell ill at a time Chow was away, she took her revenge, keeping him in the most uncomfortable accommodation with very little food or water. Next time Newton went with Chow, who by now had become suspicious and fell for an untrue tale that he was being cheated by his employee. Subsequently Newton was kept under lock and key whenever Chow was away. In desperation, John seized a chance to dispatch a letter to his father telling of his plight.

Another trader thought Chow's treatment of Newton inexcusable and offered to employ him were he released. Though his new master thought John's language was blasphemous and unfitting for a respectable slave trader. he was quickly trusted and sent, with another white man, to manage a factory a hundred miles distant. Life at last was pleasant: if only Mary were there to share it, this could be as acceptable a life-style as any.

The arrival of the *Greyhound* broke in to this new routine. Its captain was seeking the whereabouts of a John Newton. His father had offered a reward to any who obtained his release and brought him home. At first, he was reluctant to go — if he stayed another two years he might accumulate sufficient funds to be able to marry. The Captain however did not wish to forego the reward and bribed him to accept the passage home with an untrue story of a relative's legacy. Within an hour, he was a passenger on the *Greyhound.* A year later, he was in fact over a thousand miles further from London than when he boarded the vessel. During the long voyage, Newton drank heavily, and delighted in twisting the words of Scripture discovered at his mother's knee to make them sound utterly ridiculous. Indeed, even the hardened Captain of the *Greyhound* had never heard such language and feared the wrath of God might be brought down upon them all. Eventually they left Cape Lopez for home.

. . . sometimes I would promote a drinking-bout, for a frolic sake, as I termed it; for though I did not love the liquor, I was sold to do iniquity, and delighted in mischief.

John Newton: From the seventh letter of 'An authentic Narrative': 'Voyage from Cape Lopez to England'

During the long trip, the captain's fear of the wrath of God was fulfilled. *But now,* wrote Newton, *the Lord's time was come.* Disaster struck. Newton was woken by a huge wave breaking the timbers and cries of *She's sinking.* With beeswax and wool both lighter than water, Newton raided the cargo for materials to use to stuff the leaks as battle was fought for

I thought I saw the hand of God displayed in our favour; I began to pray. – I could not utter the prayer of faith; I could not draw near to a reconciled God, and call him Father. My prayer was like the cry of the ravens . . .

John Newton: From the eighth letter of 'An authentic Narrative': 'Danger etc. in the Voyage from Cape Lopez'

the ship's survival. Although exhausted, he lashed himself to the wheel and steered the ship for eleven hours. Then he began to pray. The incident is caught up in the second line of *Amazing grace! (how sweet the sound)*. Grace is to be a recurring theme when the seaman turns to hymn-writing in later life.

Amazing grace! (how sweet the sound)
That sav'd a wretch like me!
I once was lost, but now am found,
Was blind, but now I see.

'Twas grace that taught my heart to fear,
And grace my fears reliev'd;
How precious did that grace appear,
The hour I first believ'd!

Thro' many dangers, toils and snares,
I have already come;
'Tis grace has brought me safe thus far,
And grace will lead me home.

The Lord has promis'd good to me,
His word my hope secures;
He will my shield and portion be,
As long as life endures.

Yes, when this flesh and heart shall fail,
And mortal life shall cease;
I shall possess, within the vail,
A life of joy and peace.

The earth shall soon dissolve like snow,
The sun forbear to shine;
But God, who call'd me here below,
Will be for ever mine.

John Newton: 'Faith's Review and Expectation'
Sung to NEW BRITAIN (AMAZING GRACE)

. . . in the sight of God there are but two sorts of characters upon earth – the children of his kingdom, and the children of the wicked one.

John Newton: From a letter of 22nd February 1776

There were those aboard who wanted Newton thrown into the sea: they believed the tempest was God's punishment for the profanities he had uttered. But the captain resisted their demands. Provisions now grew very short, with half a salted cod a day's subsistence for twelve persons. There was sufficiency of fresh water but no strong liquor, no bread and hardly any clothing to keep out the icy weather. There was constant hard labour to keep the ship afloat: one man died under the combination of heavy exertions on little food. Eventually, just ahead of another storm, the ship limped into Lough Swilly in Ireland, where emergency repairs were undertaken.

From here the prodigal son wrote home. His father had given him up as dead and was about to leave England to take up the governorship of Fort York in Hudson's Bay. His letter asked, too, of Mary – and John's father visited Chatham and let it be known that if Mary should wish to marry John, the union would have his blessing. Meanwhile Newton found excellent lodgings and quickly recovered health and strength. He said prayers in church twice each day, signified to the local minister a desire to receive the holy sacrament and *engaged to be the Lord's* for ever and only his. He was nearly killed in an accident whilst out shooting and remembered how near we all may be to death and the time to give God an account of our lives. Then, the ship patched up, he sailed on to Liverpool where he met Manesty who offered him the captaincy of a slave ship, quite an undertaking for a young man of twenty-five.

By the time he reached London, his father had sailed: and they were never to meet again. He was extremely hesitant and shy in a meeting with Mary and could not pluck up courage to ask her to wed. He returned to Liverpool to serve, not as captain, but as first mate of the *Brownlow*, feeling himself too inexperienced to accept the captaincy. He wrote of his love for Mary in a long letter, saying on paper what he could not bring himself to say face to face, and received a cautious but not too unsatisfactory reply which at least confirmed that Mary at this point did not have another in view as her husband.

The eldest (as I understood some years afterwards) had been often considered by her mother and mine, as a future wife for me, from the time of her birth.
John Newton writes of a first meeting with the girl who was to become his bride in the third letter of 'An authentic Narrative' : 'Journey to Kent'

John Newton's conversion did not bring immediately full amendment of life. *I had learned to pray, I set some value upon the word of God, and was no longer a libertine, but my soul still cleaved to the dust . . . by the time we arrived in Guinea, I seemed to have forgotten all the Lord's mercies and my own engagements, and was (profaneness excepted) almost as bad as before.* On arrival in West Africa, his duties were to sail from places in the long-boat to purchase slaves. While the ship was in Sierra Leone, he found himself at the Plantanes, where he had been held captive; but this time he was a welcome trader. He saw with delight that the lime trees he had *planted were growing tall*, and promised fruit the following year when he expected to return as captain.

The Lord was forgotten until he *again interposed to save me. He visited me with a violent fever, which broke the fatal chain, and once more brought me to myself.*

My business in the long-boat, during eight months we were upon the Coast, exposed me to innumerable dangers and perils, from burning suns and chilling dews, winds, rains, and thunder-storms, in the open boat; and on shore, from long journeys through the woods, and the temper of the natives, who are in many places, cruel, treacherous, and watching opportunities for mischief. Several boats in the same time were cut off; several white men poisoned; and in my own boat I buried six or seven people with fevers. When going on shore, or returning from it, in their little canoes, I have been more than once or twice overset by the violence of the surf, or break of the sea, and brought to land half-dead (for I could not swim) . . .

John Newton recalls eight months collecting slaves in 'Voyage to Africa' from 'An authentic Narrative'

Philosophers have por'd in vain,
And guess'd, from age to age;
For reason's eye could ne'er attain
To understand a page.

Tho' to each star they give a name,
Its size and motions teach;
The truths which all the stars proclaim,
Their wisdom cannot reach.

John Newton: From 'The Book of Creation' which begins 'The book of nature open lies'

His call we obey
Like Abra'm of old,
Not knowing our way,
But faith makes us bold;
For tho' we are strangers
We have a good Guide,
And trust in all dangers,
The Lord will provide.

John Newton: From 'The Lord will provide' which begins 'Tho' troubles assail'

The ship set sail with its human cargo t Antigua and Charlestown in Sout Carolina. Newton describes his conduc there as *very inconsistent*. He *used t retire into the woods and fields*, whic he described as *favourite oratorie* to pray to and praise God, but stil frequently spent *the evening in vai and worthless living*. On the way home the *Brownlow* called at Jamaica. A lette from the Catletts awaited him. Its word of encouragement warmed his heart next time he would propose to Mar face to face. When they did meet, sh resisted at first: but eventually they wer married in Rochester on 12th Februar 1750. Three rapturous months followe as the newly-weds stayed in the Catlett' home. John bought one ticket afte another in the State Lottery, hoping tha good luck would defer separation fro Mary. His numbers did not come up an after three months his money was gon He set off for Liverpool where he wa taken on to captain the *Duke of Argyl*

Newton's log tells its own story:

Saturday 11th August Close weather, fresh gales between the South and West. At noon cast from the pier head at Liverpool, run under the topsails against the flood, at 3 p.m. anchored at Black Rock with the B.B. Upon the ebb, mored with the sheet anchor, hoisted in the longboat and punt. Came down and anchored with us the *Lamb*, Job Lewis, for New England. In the evening small rain.

Fryday 21st September . . . At 4 p.m. made the land right ahead, proved the island Grand Canaries; soon after saw the peak of Tenariff, W. ½S., a great distance, reckoned about 25 leagues . . . By a good observation I find that my octant agrees very well with the latitude laid down in the tables in the *Mariners' Compass*. If the longitude in the charts and tables are anything near the truth, we must have come between Madera and Port Sancto, tho think it strange that we saw neither. I am consequently not less than 50 leagues to the eastward of my reckoning, which as we have had constant fair winds and weather and frequent observations, must be owing to a strong current setting to the eastward, which cannot suppose less than 20 miles *per diem* from the time we passed the parallel of Cape St Vincent or 37° .

Tuesday 23rd October Fair weather, land and sea brease. In the afternoon, steering for the Bonanoes. Fell in with the *Cornwall*, Duncan. I went on board him in the punt, and he told me that the white men are all gone from the Bonanoes, upon which I bore away with him for Sierra Leon.

I have a good horse and a good road, and pretty good spirits likewise, considering that the more haste I make, the more I increase my distance from you.
John Newton, writing to his wife from St. Albans on 19th May 1750

Since you have kindly promised to write by every post, I wish we had a post every day.
John Newton, writing to his wife from Liverpool on 27th May 1750

It is some time since I saw the north star. When I am at sea, I shall watch it, at the hour we agreed upon, that I may have the pleasure of thinking that sometimes our eyes and thoughts are fixed upon the same object.
John Newton, writing to his wife from Liverpool on 20th July 1750

John Newton

Wednesday 24th October . . . all the ship's company to a man complained that the Boatswain had behaved very turbulently, and used them ill, to the hindrance of the ship's business. Having passed by several of the like offences before, I thought it most proper to put him in irons, *in terrorem*, being apprehensive he might occasion disturbance, when we get slaves on board.

We are accustomed from our infancy to call evil good, and good evil.
John Newton: From 'A Review of Ecclesiastical History'

Thursday 25th October In the morning I went down on board Mr. Langton's shallop in the White man's bay to view some slaves; was shown 7, out which I picked 3, viz: 2 men and 1 woman, brought them up and payed the goods for them.

Saturday 27th October . . . Dismissed the boatswain from his confinement, upon his submission and promise of amendment.

Monday 29th October This day, however, I bought a fine man boy from Mr. Langton, but the bars came excessive dear. I likewise went on board the French vessels, who offered some fine slaves to sell for basts, nicanners, iron and knives, but was not able to come up to their prices, for they understood little of the trade, and are consequently extremely positive.

Thursday 1st November Gave two of my gentlemen a good caning and put one (William Lees) in irons, both for his behaviour in the boat and likewise being very troublesome last night, refusing to keep his watch and threatening the boatswain.

Wednesday 7th November Beleive I have lost the purchase of more than 10 slaves for want of all the commanding articles of beer and cyder . . .

Sunday 18th November This morning bought a manboy and a girl slave . . . Will Lees attempted to leave me and hide himself on the island, and when I found him, he was very abusive, being drunk. Was obliged to give the blacks a gallon of brandy to secure him for me in irons, then took him into the boat and am determined to deliver him up to the first man-of-war I can meet . . .

Thursday 22nd November At 9 a.m. set out in the punt with Will Lees in irons to deliver him to the *Surprize* man-of-war, at Sierra Leon.

Sunday 25th November Fair weather. In the afternoon put 3 more men on board the *Surprize*, Thomas Creed, Owen Cavanagh and Thomas True, who all refused to sign receipts for their wages; delivered the bills for the balance due to them to Captain Baird and took his receipt.

Wednesday 28th November This morning at daylight had the agreeable sight of my longboat, and soon after she came on board with every body well, and brought 11 slaves, viz. 3 men, 1 woman, 2 men boys, 1 boy (4 foot), 1 boy and 3 girls (undersize), which makes our number 26, and likewise about a ton and half of Camwood.

Fryday 7th December . . . This day fixed 4 swivel blunderbusses in the barricado, which with the 2 carriage guns we put thro' at the Bonanoes, make a formidable appearance upon the main deck, and will, I hope, be sufficient to intimidate the slaves from any thoughts of an insurrection. . . .

Wednesday 9th January The traders came on board with the owner of the slave; paid the excessive price of 86 bars which is near 12£ sterling . . . But a fine man slave, now there are so many competitors, is near double the price it was formerly . . . This day buried a fine woman slave, No. 11 . . . Had some more rice brought off, to which with what came yesterday, amounts to about 1600 lb, and some fowls etc. Canoo brought off a load of water.

Fryday 11th January At 2 a.m. departed this life Andrew Corrigal, our carpenter, having been 10 days ill of a nervous fever; buried him at daylight, the 3rd in 3 weeks, and we now have 4 very ill.

Sunday 20th January A little before midnight, departed this life Mr. John Bridson, my chief mate, after sustaining the most violent fever I have ever seen 3 days . . . I am afraid his death will considerably retard our trade . . .

Fryday 25th January Yellow Will brought me a woman slave, but being long breasted and ill made, refused her . . . He brought of a cask of palm-oyl I gave him to fill . . .

Sunday 21st April Was obliged to wave the consideration of the day* for the first and, I hope, the last time of the voyage, the season advancing fast and, I am afraid, sickness too; for we have almost every day one or two more taken with a flux,

* the religious service for the crew

Most of my letters to you remind me of Aesop's feast, which, though consisting of several dishes, were all tongues, only dressed in different ways. Thus, whether I write in a grave or a jocular strain, the subject is still love, love; which is as inseparable from my idea of you, as heat from that of fire.

John Newton writing to his wife from Shebar on 21st December 1750

It was an expression of Cato, that it was more honourable to be a good husband, than a great senator.

John Newton writing to his wife from Shebar on 5th March 1751

I give and take a good deal of raillery among the sea captains I meet with here. They *think* I have not a right notion of life, and I *am sure* they have not. They say I am melancholy; I tell them they are mad. They say, I am a slave to one woman, which I deny; but can prove that some of them are mere slaves to a hundred.

John Newton writing to his wife from Shebar on 29th March 1751

Indeed, your idea is constantly with me, and I hope in due time I shall prove the reverse of Aesop's dog, and by long gaping after the shadow, come at length to repossess the substance.

John Newton writing to his wife on 27th June 1751 whilst at sea.

. . . I should have been afraid of a hurricane (for this is the season), but that my dependence upon the providence of God is become almost habitual.

John Newton writing to his wife on 19th August 1751 whilst at sea.

I am a great admirer of Aesop's fables. They could hardly have been more adapted to the customs and humours of our times, had they been written in London. His apes, lions, foxes, geese, magpies, and monkeys, may be in our streets every day. As a proof that I am not partial in my censure, I will confess that I myself have frequently appeared in some of these characters. When I first knew you, I was a bear; I then became an owl, and afterwards exhibited the worst properties of *all* his brutes in my single self.

John Newton writing to his wife on 19th September 1751 whilst at sea.

. . . that I may not be like the sailor who once, in great distress, made a vow to the Virgin Mary, that if she would deliver him, he would present her with a wax-candle as big as the ship's main-mast; and, on being asked how he would raise money to pay for so large a candle, he said, ***Let us first get on shore, and then the saints will not exact too strictly upon a sailor's promise.***

John Newton writing to his wife on 2nd October 1751 whilst at sea.

of which a woman dyed tonight (No. 79). I imputed it to the English provisions and have given them rice twice a day ever since I came here; a little time will show whether it agrees better with them than beans or pease.

Tuesday 23rd April . . . got on board about 3800 lb. of rice . . .

Fryday 24th May Fair pleasant weather, light airs and calms. . . . Sounded several times for fear that I should have been unexpectedly drawn too near the Shoals of St. Anne.

Sunday 26th May In the evening, by the favour of Providence, discovered a conspiracy among the men slaves to rise upon us, but a few hours before it was to have been executed. A young man, No. .., who has been the whole voyage out of irons, first on account of a large ulcer, and since for his seeming good behaviour, gave them a large marline spike down the gratings, but was happily seen by one of the people. They had it in possession about an hour before I made search for it, in which time they made such good dispatch (being an instrument that made no noise) that this morning I've found near 20 of them had broke their irons. Are at work securing them.

Monday 27th May . . . A hard tornado came on so quick that had hardly time to take in a small sail; blew extream hard for 3 hours with heavy rain . . . punished 6 of the ringleaders of the insurrection.

Saturday 22nd June In the fore noon, being pretty warm, got up the men and washed all the slaves with fresh water. I am much afraid of another ravage from the flux, for we have had 8 taken within these last few days. Have seen 2 or 3 tropick birds and a few flying fish.

Saturday 29th June Have seen several men of war, boobies and other fowls and frequent flocks of small birds. Have likewise a good many flying fish about us, but no gulphweed. No appearance of land. Buryed a man slave (No. 2) of a flux sustained about 3 months. Washed the slaves with fresh water.

So the *Duke of Argyle* brought her human cargo to Antigua. On 13th August, Newton set sail for Liverpool. Within four days, the thirty-fifth death of the trip was of Robert Arthur the surgeon. All told, seven whites died and twenty-eigh of one hundred and seventy four blacks, a below average roll for a British slave ship of the period. On Monday, 7th October, the journal recalls that *at 4 a.m. got a pilot on board, worked up the flood to Lyle-lake, and anchored at 10 just within the buoy.* It took a few long days for the business details at the end of the voyage to be settled; then, at top speed, John made for Chatham and Mary.

The winter passed all too fast and early in the spring, John returned to Liverpool, and saw the new ship *African* he was to captain launched on 19th April. He was to describe it as one of the strongest vessels that could be bought for money, but *a very* indifferent sailor. His diary for 5th July, 1752, describes the first Sunday at sea. *This day I have reason to be particularly observant of my Duty in keeping the Sabbath holy to the Lord both as it is the first opportunity I have had of convening my ship's company to beg a publick blessing from Almighty God upon our voyage; and farther as it is a day appointed . . . for celebrating the sacrament of the body and blood of my Lord Jesus, in which communion tho' I cannot personally join yet my heart and thoughts ought to be engaged upon so interesting a theme.*

On these occasions, all the people on board a ship who have not passed it before, are subject to a fine, which, if they refuse to pay, or cannot procure, they must be ducked; that is, hoisted up by a rope to the yard-arm, and from thence dropped souse into the water. This is such fine sport to the seamen, they they would rather lose some of the forfeiture (which is usually paid in brandy) than that every body should escape the ducking . . . But as I do not choose to permit any arbitrary or oppressive laws to be valid in my peaceful kingdom, I always pay for those who cannot pay for themselves.

John Newton writing to his wife on 27th July 1752 about the ceremony of crossing the line observed at the tropic of Cancer.

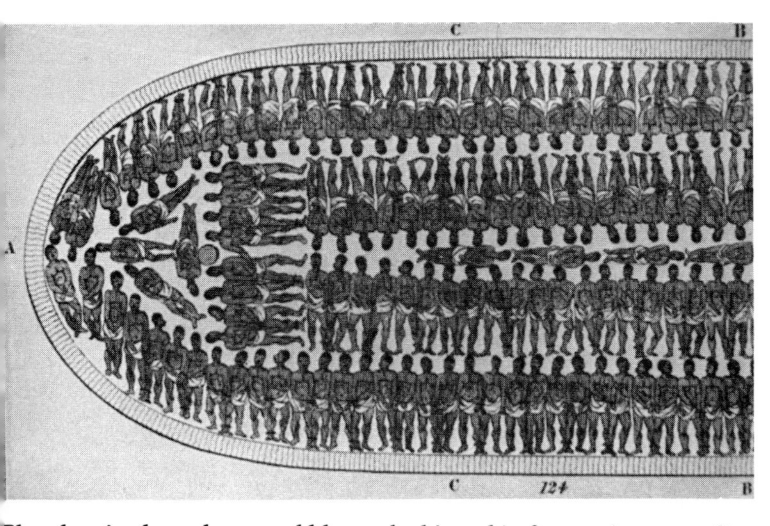

Plan showing how slaves could be packed in a ship for maximum profit

Begone unbelief,
My Saviour is near,
And for my relief
Will surely appear:
By pray'r let me wrestle,
And he will perform,
With Christ in the vessel,
I smile at the storm.

Tho' dark be my way,
Since he is my guide,
'Tis mine to obey,
'Tis his to provide;
Tho' cisterns be broken,
And creatures all fail,
The word he has spoken
Shall surely prevail.

His love in time past
Forbids me to think
He'll leave me at last
In trouble to sink;
Each sweet Ebenezer
I have in review,
Confirms his good pleasure
To help me quite thro'.

Determin'd to save,
He watch'd o'r my path,
When Satan's blind slave,
I sported with death;
And can he have taught me
To trust in his name,
And thus far have brought me,
To put me to shame?

Why should I complain
Of want or distress,
Temptation or pain?
He told me no less:
The heirs of salvation,
I know from his word,
Thro' much tribulation
Must follow their Lord.

How bitter that cup,
No heart can conceive,
Which he drank quite up,
That sinners might live!
His way was much rougher,
And darker than mine;
Did Jesus thus suffer,
And shall I repine?

Since all that I meet
Shall work for my good,
The bitter is sweet,
The med'cine is food;
Tho' painful at present,
Will cease before long,
And then, oh! how pleasant,
The conqueror's song!

John Newton: 'I will trust and not be afraid'
OLD 104th (RAVENSCROFT);
LAUDATE DOMINUM (Gauntlett)
(GAUNTLETT); SPETISBURY

The three greatest blessings of which human nature is capable, are undoubtedly, religion, liberty, and love. In each of these how highly has God distinguished me!
John Newton writing to his wife from Mana on 26th January 1753

Newton then was still unsure about his fitness to be called a Christian. He resolved however to begin every morning with one hour in prayer and reading the Bible, or if not otherwise hindered, two hours. After that, if time allowed, he improved his knowledge of Latin, French and Mathematics. He did his duty in reading the words of the Book of Common Prayer to the crew: but did not find them always especially appropriate to the life of the seamen.

On his return to Liverpool, he wrote to Dr. David Jennings pleading that someone would write and publish a book of more suitable prayers for such occasions. His journal tells of a plot by at least three of his crew to take over the ship. He was surprised by this as he believed he had a contented crew. Someone informed on the renegades: one was struck down with a fever and died, two others were transferred to another ship to be handed over to the Royal Navy. Later, after his cargo of slaves had been rounded up, he found four of them planning insurrection. He punished them with the humbscrews and afterwards put them in neck yokes. His diary later in the voyage gives thanks to the God of peace for a remarkable change in the demeanour of his human cargo. They behaved more like children in one family than slaves in chains and irons. They were really on all accounts more observant, obliging and considerate than white people! The last of this family, numbering one hundred and sixty even, was sold on Wednesday, 20th June 1753 and the details entered in the ship's accounts. The price was of pecial concern to John Newton: a substantial part of his emuneration was calculated on a profit-sharing basis.

How could I bear that you should love me less than you do!
John Newton writing to his wife from Shebar on 5th March 1753

Most of the cargo is sold, and at a good price.
John Newton writing to his wife from Sandy Point, St. Kitts on 8th June 1753

At this point of time, Newton still saw no conflict with his occupation as master of a slave ship and his striving towards a Christian calling, save that there was a responsibility not o treat the cargo more harshly than necessary compatible with their being duly delivered for sale. On the other hand, he was concerned about the soul of his brother-in-law, making good progress in his calling as a lawyer but ndifferent to the demands of religion. In une 1753, he wrote from St. Kitts to ohn Catlett: *I look upon your opinions as dangerous to profess and destruction o persist in. As to what you say of the mpossibility of forcing ourselves to believe, I must call it downright ophistry. I acknowledge that belief s not immediately in our power but the means are.* He continued to point out he folly and danger of infidelity hrough the years until John Catlett's death in 1764.

Then the rich, the great, the wise,
Trembling, guilty, self-condemn'd;
Must behold the wrathful eyes
Of the Judge they once blasphem'd:
Where are now their haughty looks?
Oh their horror and despair!
When they see the open'd books
And their dreadful sentence hear!

– – – – –

Oh! when flesh and heart shall fail,
Let thy love our spirits cheer;
Strengthen'd, we shall prevail
Over Satan, sin, and fear:
Trusting in thy precious name,
May we thus our journey end;
Then our foes shall lose their aim,
And the Judge will be our Friend.

John Newton: The third and final verses of 'Prepare to meet God' which begins 'Sinner, art thou still secure?'

He was hardly back than the *African* was off again on another voyage. The log is little different from that of the previous rip. He took a tough line with the crew *correcting the*

I have been ill three days of a fever, which, though it is at present attended with no symptoms particularly dangerous, it behoves me to consider may terminate in death.

John Newton writing to his wife on 18th April 1754 whilst at sea.

. . . I thought it right to consider it as a warning to prepare for eternity. . .

John Newton writing to his wife on 16th May 1754 whilst at sea.

I usually rise at four on a Sunday morning. My first employ is to beg a blessing upon the day for us both; for all who, like you, are preparing to wait upon God in public, and for all who, like myself, are, for a time, excluded from that privilege. To this succeeds a serious walk upon deck. Then I read two or three select chapters. At break--fast, I eat and drink more than I talk; for I have no one here to join in such conversation as I should then choose. At the hour of your going to church, I attend you in my mind with another prayer; and at eleven o'clock the ship's bell rings my own little congregation about me. To them I read the morning service, according to the Liturgy.

John Newton writing to his wife on 27th July 1754 whilst at sea.

carpenter with a catt – gave him 2 dozen stripes on 21st December 1753, only sparing him confinement in irons because there was too much joinery to be done before slaves were loaded. The trip was however unique in that only one life was lost, of free man or slave, during the trip, and this not of a paid member of the crew. Just before sailing, he met Job Lewis, a former midshipman from his days aboard *HMS Harwich*, who had just been prevented from sailing as a captain by the bankruptcy of his employer. Newton took him along as a volunteer deputy but soon regretted his decision. The man was an unreformed reprobate, and when he died of a fever in the depths of despair, religion was once more pushed uppermost in John's mind. He had of course little contact with other committed Christians during the long sea voyages but talked at length with Captain Alexande Clunie at St. Kitts. Clunie urged on him the need and duty of a public profession of belief and directed him to the Reverend Samuel Brewer, Minister of the Independent Church in Stepney from 1746 to 1796.

He had earned a reputation as an excellent captain an Joseph Manesty appointed him to a new and better shi *The Bee*. Two days before he was due to sail, Captain Joh Newton was sharing tea with his wife when he collapse in some kind of fit. After an hour, when he showed no sig of life, he regained consciousness: but sailing in comman of *The Bee* was clearly out of the question. As a youn man without other means of livelihood, this was a real worry perhaps more for Mary than for John who was also minde to use every opportunity to relate to others how he was living example of how a sinner could be saved. Mary becam quite seriously ill and none of the treatments offered by th doctors seemed to be of any avail.

The couple returned to Chatham and whilst in the South, h took Captain Clunie's advice and attended Mr. Brewer' church and consulted with him. He was also introduced by Mr. Hayward to the evangelist, George Whitefield. Meanwhil in the matter of employment, Joseph Manesty again came t the rescue. He used his good offices to secure for Newto the post of one of Liverpool's two tide surveyors. Now a last, John had a shore job: alas, Mary was still too ill to trave back with him to Liverpool.

When off duty, John became involved with local Baptists an Dissenters and moved towards their quieter life style. He als became convinced that his love of money made him prone t the temptation of accepting bribes from those trying t smuggle contraband through the port: his conscience forced him to give up this lucrative practice. He worshipped his wife and he even wondered if this was acceptable to the God who demanded total allegiance. In the hymn he wrote later *Come, my soul, thy suit prepare*, he ends the fourth verse *And without a rival reign.* Eventually though he was delighted to write: . . . *the Lord was pleased to restore her by his own hand, when all hopes from ordinary means were at an end.*

In 1757, Newton began frequent visits to the West Riding of Yorkshire, describing it *as a good school*, a place *where the Gospel flourishes greatly*. He mixed with Christians of *all parties without joining any*. He resolved to devote his *life to the pursuit of spiritual knowledge.* He gave up any interest in the classics and mathematics, aiming to learn as much Greek to understand the New Testament and Septuagint and over the next years went on to study Hebrew and Syriac that he might better understand the Scriptures. He made it clear he had no wish to be an eminent linguist – these studies were essentially a means towards his primary purpose. During his visits to Yorkshire, he met several Christian ladies with whom he began a correspondence which lasted several years. One such was a Miss Medhurst, a relative of the Countess of Huntingdon. Also at this period, he came to know a Mr. Whitford, then connected with the Methodists and they became close friends.

He describes his first experience as a preacher at a dissenting meeting-house in Leeds. He was determined to speak extempore. He had his subject divided under four heads, with subdivisions in mind for each and began well enough. Then his mind went completely blank. *I was stopped, like Hannibal upon the Alps. My ideas forsook me; darkness and confusion filled up their place. I stood on a precipice, and could not advance a step forward. I stared at the people, and they at me. But I remained as silent as Friar*

Come, my soul, thy suit prepare,
Jesus loves to answer pray'r;
He himself has bid thee pray,
Therefore will not say thee nay.

Thou art coming to a King,
Large petitions with thee bring;
For his grace and pow'r are such,
None can ever ask too much.

With my burden I begin,
Lord, remove this load of sin!
Let thy blood, for sinners spilt,
Set my conscience free from guilt.

Lord! I come to thee for rest,
Take possession of my breast;
There thy blood-bought right maintain,
And without a rival reign.

As the image in the glass
Answers the beholder's face;
Thus unto my heart appear,
Print thine own resemblance there.

While I am a pilgrim here,
Let thy love my spirit cheer;
As my Guide, my Guard, my Friend,
Lead me to my journey's end.

Shew me what I have to do,
Ev'ry hour my strength renew;
Let me live a life of faith,
Let me die thy peoples death.

John Newton: 'Ask what I shall give thee'
Tunes include CANTERBURY (SONG 13) (SIMPLICITY (Gibbons)); LOUEZ DIEU (PSALM 136); RICHMOND (Stephens); SAVANNAH (HERRNHUT (Wesley)); THEODORA (Handel); TUNBRIDGE

As there are two surveyors, and I shall be upon the river only every other week, the place is likely to afford me leisure, which, in its turn, will be as welcome to me as money.
John Newton writing to his wife from Liverpool on 15th August 1755

I have a good office, with fire and candle, fifty or sixty people under my direction, with a handsome six-oared boat and a coxswain, to row me about in form.

John Newton writing to his wife from Liverpool on 20th August 1755

Bacon's head, and was forced to come down re infecta. *My two worst enemies, Self and Satan, seized me at the bottom of the stairs.*

At least he did not give up the thought of becoming a preacher, though he wondered about it. He went instead to the other extreme, writing out his discourse to the very last word and keeping his eyes glued to his notes whilst delivering a sermon, which consequently proved an ineffective communication.

At that time, he began to invite a few friends home on Lord's Day evenings and continued the practice for the rest of his residence at Liverpool. In this situation, *God was pleased in some measure to open* his *mouth.* Many years later Newton set down a possible plan of academical preparation for the ministry. More immediately he himself became determined that he should eventually be ordained.

1 An orderly, connected, and comprehensive knowledge of the common places and topics of divinity, considered as a whole; a system of truth, of which the holy Scripture is the sole fountain, treasury, and standard.

2 A competent acquaintance with sacred literature; by which I mean such writings, ancient and modern, as are helpful to explain or elucidate difficulties in Scripture arising from phraseology, from allusion to customs and events not generally known, and from similar causes, and which therefore cannot be well understood without such assistance.

3 Such a general knowledge of philosophy, history, and other branches of polite literature, as may increase the stock of their ideas, afford them just conceptions of the state of things around them, furnish them with a fund for variety, enlargement, and illustration, that they may be able to enliven and diversify their discourses, which, without such a fund, will be soon apt to run in a beaten track, and to contain little more than a repetition of the same leading thoughts, without originality or spirit.

4 An ability to methodize, combine, distinguish, and distribute the ideas thus collected by study, so as readily to know what is properly adapted to the several subjects to be treated of, and to the several parts of the same subject. When the pupils are thus far accomplished, then I shall hope:

Let us not be greatly discouraged at the many tribulations, difficulties, and disappointments which lie in the path that leads to glory: seeing our Lord has told us before; has made a suitable provision for every case we can meet with; and is himself always near to those that call upon him; a sure refuge, an almighty strength, a never-failing, ever-present help in every time of trouble; seeing likewise that he himself was a man of sorrow, and acquainted with grief for our sakes.

John Newton: From a letter to Miss Medhurst written on 2nd November 1761

5 That they will in good time be able to preach extempore. I do not mean without forethought or plan, but without a book, and without the excessive labour of committing their discourses to memory. This ability of speaking to an auditory in a pertinent and collected manner, with freedom and decorum, with fidelity and tenderness, looking at them instead of looking at a Paper, gives a preacher a considerable advantage, and has a peculiar tendency to command and engage the attention. It likewise saves much time, which might be usefully employed in visiting his people. It is undoubtedly a gift of God, but, like many other gifts, to be sought not only by prayer, but in the use of means. The first essays will ordinarily be weak and imperfect; but the facility increases, till at length a habit is formed, by diligence and perseverance. I should not think my academy complete, unless my tutor was attentive to form his pupils to the character of public speakers.

John Newton: From 'A Plan of Academical Preparation for the Ministry'

We live in dangerous times; the work of the Lord is greatly on the revival in many places, and therefore errors and offences abound; for where the good seed is plentifully sown, the enemy will always find means to sow his tares.

John Newton: From a letter to Miss Medhurst written on 22nd March 1763

)n 3rd April 1759, Newton initiated a correspondence with a 1oravian, the Reverend Francis Okeley. He described him as .nowing *not if* his *heart was ever more united to any person, ı so short a space of time, than to* him: having observed in .im *the spirit of meekness and of love (that peculiar and ıimitable mark of true Christianity).* Francis Okeley had ejected entry into the Church of England because he upposed this would restrain him from preaching his under-tanding of the Gospel. Newton's first letter assures him of he love of many in Yorkshire he had left rather hurriedly. Ie worked with the Reverend B. Ingham and later, with the Reverend Jacob Rogers promoting the truth in Bedfordshire. ıfter this, he served as a Moravian minister in Northampton.

Iewton's own attempts to become a parson were full of rustration. His in-laws were against it, thinking it foolish to ive up the safe job he had at Liverpool. He had determined e wished to serve in the Church of England, but his many ontacts with Dissenters made him suspect in this regard. Iis formal education had been scanty and, as John Wesley old him, an intending minister should have had a university ducation. Why, anyway should the Church of England wish ɔ ordain a former sea-captain, now in his thirties, even if he ad showed a mental courage in his home studies in trying to ıatch his valour in seamanship with a mastery of Greek, ,atin, Hebrew and Syriac. The Bishop of Chester refused. 'he Archbishop of York refused. He was offered a chance of ecoming the minister of a Presbyterian Church in Yorkshire 'hen Lord Dartmouth intervened.

I loved liberty, and therefore gave preference to the church of England, believing I might in that situation exercise my ministry with the most freedom.

John Newton on reasons for not being a dissenter in 'Apologia: Four letters to A Minister of an Independent Church by A Minister of the Church of England'

John Wesley

Mrs. Newton has kept her chamber more than ten weeks; and we see no present prospect of her recovery. Her complaint is a nervous fever, attended with a complaint in her head and stomach, which medicines seem insufficient to remove.

John Newton in a letter to Miss Medhurst dated 3rd January 1764

The story of Newton's life at sea and conversion had beer put on paper in a series of letters written at the recipient's request to the Reverend T. Haweis, and together formec an autobiography which Newton published in 1764. Lorc Dartmouth, whom William Cowper described as *the pee who wears a coronet and prays*, was one of the few in higl places in sympathy with the evangelical revival. Later he became President of the Board of Trade, Secretary of State for the Colonies, Lord Keeper of the Privy Seal and Lorc Steward of His Majesty's Household. Lord Dartmouth wa greatly impressed by Newton's story which Mr. Haweis hac shown him and had in his gift the living of Olney. The Vica there was the Reverend Moses Brown who had a very large family but the small stipend was insufficient to maintair them. So, Lord Dartmouth arranged for Moses Brown whilst remaining non-resident vicar of the lace-making town to accept the chaplaincy of Morden College, Blackheath. He was quite a versifier in his own right and was at one time a substantial contributor to *The Gentleman's Magazine*.

It has been ingeniously (not without shew of reason) supposed, that Adam was created on Friday, at 3 in the afternoon, and fell on Sunday morning, else why that hour (viz. of 3) for the daily sacrifice.

Moses Browne: Footnote to 'Sunday Thoughts'

Oft from the blighting East the Locust bands,
In swarthy numbers, shade the vernal lands.
The naked fields lament the wasteful drove,
The ruin'd harvest, and the leafless grove.
And oft the South malignant seasons brings,
And sickly Autumns on her burning wings:
When livid poisons taint her stagnant breath,
With putrid streams, and pestilential death.

Moses Browne: From 'An Essay on the Universe'

Lord Dartmouth lined up the curacy a Olney for John Newton and persuade the Bishop of Lincoln to ordain him Before the Newtons took up residence i their new home, John preached six time in Liverpool to congregations totallin several thousands. God had overcom the inability to preach! Newton, too was offered the better paid living o Hampstead, then just outside London but refused it, being certain God had called him to a ministr among the poor lace makers. The church, probably built i the fourteenth century, stands near the Great Ouse River Today, a bird sanctuary separates the church from the bank The church spire is built with an entasis and this bulge add to its majestic appearance.

So there began a five-fold ministry in Olney and later i London. He was pastor to his parish; he carried on a ministr by correspondence to a large circle; he published sermons an other writing; he wrote hymns; and God gave him a specia part to play in the lives of others called to special ministrie:

However, with all their faults, methinks there is something very beautiful and engaging in the honest vehemence of a young convert. Some cold and rigid judges are ready to reject these promising appearances on account of incidental blemishes. But would a gardener throw away a fine nectarine, because it is *green*, and has not yet attained all that beauty and flavour which a few more showers and suns will impart?

John Newton: From a letter to Lord Dartmouth written in March 1765

John Newton threw himself into the parish work, making re contact with the poverty-stricken lace-makers: there wer complaints that he let down the worthiness of his office b his own familiarity with parishioners. More than half th population were paupers and children ran wild in the streets

with the help of Lord Dartmouth, he managed to start a Sunday School. He was also helped to alleviate some of the worst poverty with funds given to him to be used at his own discretion by John Thornton, one of the richest commercial men in the country. William Cowper later wrote of Thornton:

Thou had'st an industry in doing good,
Keen as the peasant's toiling for his food.

Newton's preaching was unusual, concentrating on his own life story and his experience of personal salvation. Soon the church was packed as people came from far afield to hear the slave-captain turned priest.

Chief Shepherd of thy chosen sheep,
From death and sin set free;
May ev'ry under-shepherd keep
His eye, intent on thee!

With plenteous grace their hearts prepare,
To execute thy will;
Compassion, patience, love and care,
And faithfulness and skill.

Enflame their minds with holy zeal
Their flocks to feed and teach;
And let them live, and iet them feel
The sacred truths they preach.

Oh, never let the sheep complain
That toys, which fools amuse;
Ambition, pleasure, praise or gain,
Debase the shepherd's views.

He, that for these, forbears to feed
The souls whom Jesus loves;
Whate'er he may profess, or plead,
An idol-shepherd proves.

The sword of God shall break his arm,
A blast shall blind his eye;
His word shall have no pow'r to warm,
His gifts shall all grow dry.

Olney Parish Church

. . . humility is necessary and beautiful in a minister.

John Newton: From 'A Review of Ecclesiastical History'

O Lord, avert this heavy woe,
Let all thy shepherds say!
And grace, and strength, on each bestow,
To labour while 'tis day.

John Newton: 'Prayer for Ministers'
This continues to be sung to-day but its first line 'Chief Shepherd of thy chosen sheep' is usually altered to begin with 'Great' or 'Dear'. Tunes used include ABRIDGE (ST. STEPHEN (Smith)), OSWALD'S TREE, RENDEZ A DIEU (GENEVAN PSALM 98; GENEVAN PSALM 118), ST. FLAVIAN (FLAVIAN; OLD 132nd, ST. STEPHEN (Jones) (NEWINGTON (Jones)) and WETHERBY

A minister of Jesus Christ is as high a style (according to the spiritual heraldry in the word of God) as mortal man can attain. His department is much more important than that of a first Lord of the Treasury or Admiralty, a Chancellor, or a *mere* **Archbishop.**

John Newton, writing to Daniel West on April 1773

A gallery had to be built to accomodate the worshippers. He also initiated a regular Thursday meeting, with a considerable attendance which included many dissenters. Newton's preaching was not confined to Olney; there is a record of his preaching fourteen times on a visit to London. When at his home base, a typical day would devote the morning to writing and reading, and a couple of hours in the afternoon to visiting around the parish. His many correspondents included Mrs. Wilberforce, the sister of John Thornton and wife of William Wilberforce, the uncle of the celebrated statesman of the same name. He wrote often to the Earl of Dartmouth, this private correspondence being later published under the title *Letters to a Nobleman*. Lord Dartmouth had first been acquainted with the evangelical movement through the friends of Lady Huntingdon with whom he met at her home, including George Whitefield and the Wesleys. Another cause Lord Dartmouth was to embrace was a College for the American Indians. Indeed Lord Dartmouth wanted Newton to go to Georgia and to combine its Presidency with the living of Savanna. He turned it down with twin reasons of love for Olney and Mary Newton's hatred of the thought of sea travel.

It was intended to pay you only one visit. But the reception of the gospel met with pleaded effectually for a second, for a third, for many: till the preacher reckoned it among his chief privileges to be with you. The number of hearers increased, with almost every visit, for near, if not quite, two years: till it was found necessary to enlarge the place of our meeting. Before long, our enlarged space was found too small; and then, by a very distinguishing Providence, this place was built.

Thomas Bowman: From 'Caustonia: Discourse IX addressed to the inhabitants of the Parish of Cawston, in Norfolk.

The Reverend Thomas Bowman, the Vicar of Martham in Norfolk found his heart strangely warmed in 1763 but still found himself handicapped with considerable shyness. He sought John Newton's guidance, and after corresponding, Newton visited him in 1768. A great awakening was reported in Martham in 1774. Bowman also published a book *The Principles of Christianity as taught in Scripture*. John Wesley had initiated a correspondence in which he and Newton openly discussed differences on certain points of theology whilst confirming their unity of fellowship in the Lord Jesus Christ. A new contact made in 1765 was the Reverend John Ryland, the Baptist minister of Northampton.

George Whitefield

In January 1766, Newton wrote to Daniel West, a friend of the Reverend George Whitefield, who had left him with a Mr. Keene as joint trustees of The Tabernacle and Tottenham Court Chapel. Of Daniel West's recent troubles, the Olney curate corresponded: *You, (and consequently Mrs. West, for you cannot suffer alone,) have lately been in the furnace, and are now safely brought out. I hope you have much to say of the grace, care, and skill of the great Refiner, who watched over you; and that you have lost nothing but dross . . .* Daniel West was to suffer persistent ill-health and his wife was consequently anxious that death would snatch her husband from her. Newton told her that natural as this conern for her husband was, all the worry in the world would neither hasten nor postpone the day he would start the new life beyond death: and she must trust the Lord in all things, even in this.

Were we always alike, we should dream that we had some power or goodness inherent in ourselves; he will therefore sometimes withdraw, that we may learn our absolute dependence on him.

John Newton: From a letter to Miss Medhurst written around 1769

. . . I began to expound the *Pilgrim's Progress* in our meetings on Tuesday evenings; and though we have been almost seven months travelling with the pilgrim, we have not yet left the house Beautiful; but I believe shall set off for the Valley of Humiliation in about three weeks . . .

John Newton: From a letter to Miss Medhurst written in July 1768

It is a happy and most desirable frame to be ready and willing either to live or die, and to be enabled so absolutely to give ourselves up to the Lord's disposal as to have no choice of our own either way, but only intent upon improving to-day, and cheerfully to leave to-morrow and all beyond it in his hands who does all things well.

John Newton: From a letter written to Daniel West on 8th July 1769

On 12th March 1766, Newton noted a visit from Reverend Joshua Symonds, who was being considered as the pastor for the chapel in Bedford where John Bunyan once ministered. Newton was glad that he accepted the call to minister there, thinking *him a sensible, spiritual, humble young man* who would make *an agreeable and useful neighbour.* On 14th June, he went in a postchaise to collect the Reverend Samion Occum, a Mohegan Indian to preach. He also went on a six hundred and fifty mile preaching trip which took in Berwick Jail. On Thursday, 7th October, a day set aside for fasting, he described fine weather and was delighted to wander in the woods and fields. That same year he turned down another offer of a parish at Cottenham.

Unasked by the Newtons, Lord Dartmouth arranged for a greatly enlarged vicarage into which the family moved in 1767. Mrs. Mary Unwin and her daughter with a Mr. William Cowper who lodged with them came to spend a night at Olney. Mary was the recent widow of a clergyman who had recently been thrown from his horse and killed: she was looking for a new home in a place where there was an evangelical clergyman. They found the home at *Orchard Side* on Olney's Market Square and lodged at the Vicarage while the house was being prepared for occupation.

This was the beginning of the friendship between John and Mary Newton with Mary Unwin and William Cowper, who was to become a permanent lodger. He was just recovering from a severe bout of mental illness, a condition that was to return periodically over the years. Newton found in Cowper a willing helper in the parish work and more significantly a poet. Often Cowper retired to the vicarage for sanctuary when he could not face the rigours of daily life. It was in part to keep Cowper from despondency that Newton suggested they should both write hymns. Most of these were written either to illustrate Newton's sermons or for parish prayer meetings. Eventually the 348 hymns were published in 1779 under the title *Olney Hymns.*

My friend, I watched thee, when that earthly frame
Encircled (union strange) the ethereal flame,
And there were hidden sufferings, that no eye
Of skilled and kind physician could descry;
And there were doubts, and fears, and terrors given,
Till peace on earth was gone, and hope of heaven.

John Newton: From 'Newton watches over him' concerning his friend William Cowper during one of his manic depressions.

Another correspondent at this time was an Army officer, Captain Scott who was converted to the Gospel Faith. His commanding officer did not like his officers to take this measure of interest in religion and sought his resignation. Captain Scott wondered if he should become a minister of the Established Church, a course from which Newton dissuaded him, but as he left the Army,

Newton continued to give counsel by post and in person.

Thomas Jones was one of six students expelled in 1768 from St. Edmund's House, Oxford, *for holding Methodist tenets, and taking upon them to pray, read and expound the Scriptures, and singing hymns in private houses.* Thomas came to live with the Newtons and this former hairdresser made considerable progress in the study of the Greek and Hebrew Scriptures, later becoming curate of Clifton, a village near Olney.

In these early days of the Olney ministry, Newton completed his *A Review of Ecclesiastical History, so far as it concerns the Progress, Declensions and Revivals of Evangelical Doctrine and Practice* which was published in 1770.

It must be confessed that the bulk of Ecclesiastical History, as it is generally understood, is little more than a history of what the passions, prejudices, and interested views of men have prompted them to perpetrate, under the pretext and sanction of religion. Enough has been wrote in this way; curiosity, nay, malice itself, need desire no more. I propose to open a more pleasing prospect, to point out, by a long succession of witnesses, the native tendency and proper influence of the religion of Jesus; to produce the concurring suffrage of different ages, people, and languages, in favour of what the wisdom of the world rejects and reviles; to bring unanswerable proofs that the doctrine of grace is a doctrine according to godliness, that the constraining love of Christ is the most powerful motive to obedience, that it is the property of true faith to overcome the world, and that the true church and people of Christ have endured his cross in every age; the enemy has thrust sore at them that they might fall, but the Lord has been their refuge and support; they are placed upon a rock that cannot be shaken, they are kept guarded and garrisoned by the power of God, and therefore the gates of hell have not, cannot, shall not prevail against them.

Introduction to John Newton's 'A Review of Ecclesiastical History so far as it concerns the Progress, Declensions and Revivals of Evangelical Doctrine and Practice', 1770

This selection of texts linked to the Old Testament is taken from the first section of *Olney Hymns.*

. . . Beware of being too happy – beware of idolatry. Husbands, children, possessions, every thing by which the Lord is pleased to afford us content or pleasure, are full of snares. How hard is it to love a creature just as we ought; and so to possess our temporal blessings as neither to overvalue nor undervalue them!

John Newton: From a letter written in September 1770

The Gospel is a wise and gracious dispensation, equally suited to the necessities of man, and to the perfections of God. It proclaims relief to the miserable, and excludes none but those who exclude themselves.

John Newton: From 'A Review of Ecclesiastical History'

Perfection cannot be found in fallen man. The best are sometimes blamable, and the wisest often mistaken. Warm and active tempers, though influenced, in the main, by the noble ambition of pleasing God in all things, are apt to overshoot themselves, and to discover a resentment and keenness of spirit which cannot be wholly justified. Others of a more fixed and sedate temper, though less subject to this extreme, are prone to its opposite; their gentleness degenerates into indolence, their caution into cowardice.

John Newton: From 'A Review of Ecclesiastical History'

Such was the wicked murd'rer Cain,
And such by nature still are we,
Until by grace we're born again,
Malicious, blind and proud, as he.

John Newton: From 'Cain and Abel' which begins 'When Adam fell he quickly lost'

Kings are often waking kept,
Rack'd with cares on beds of state;
Never king like Jacob slept,
For he lay at heaven's gate:
Lo! he saw a ladder rear'd,
Reaching to the heav'nly throne;
At the top the Lord appear'd,
Spake and claim'd him for his own.

John Newton: From 'Jacob's Ladder' which begins 'If the Lord our leader be'

If Solomon for wisdom pray'd,
The Lord before had made him wise;
Else he another choice had made,
And ask'd for what the worldlings prize.

John Newton: Opening verse of second hymn entitled 'Ask what I shall give thee'

As vain was the decree
Which charg'd them not to pray;
Daniel still bow'd his knee,
And worshipp'd thrice a day:
Trusting in God, he fear'd not men,
Tho' threatened with the lion's den.

John Newton: From 'Daniel' which begins 'Supported by the word'

Poor Esau repented too late
That once he his birth-right despis'd;
And sold, for a morsel of meat,
What could not too highly be priz'd:
How great was his anguish when told,
The *blessing* he fought to obtain,
Was gone with the *birth-right* he sold,
And none could recall it again!

John Newton: Opening verse of 'Esau'

The signs which God to Gideon gave,
His holy Sov'reignty made known;
That He alone has pow'r to save,
And claims the glory as his own.

John Newton: Opening verse of 'Gideon's Fleece'

His wife escap'd a little way,
But dy'd for looking back:
Does not her case to pilgrims say,
Beware of growing slack?

John Newton: From 'Lot in Sodom' which begins 'How hurtful was the choice of Lot

Elijah's example declares,
Whatever distress may betide;
The saints may commit all their cares
To him who will surely provide:
When rain long withheld from the earth
Occasion'd a famine of bread,
The prophet, secur'd from the dearth,
By ravens was constantly fed.

John Newton: Opening verse of 'Elijah fed by Ravens

Yea, Aaron, God's anointed priest,
Who on the mount had been;
He durst prepare the idol-beast,
And lead them on to sin.

John Newton: From 'The golden Calf which begins 'When Israel heard the fiery law

When she began to pray,
Her heart was pain'd and sad;
But, ere she went away,
Was comforted and glad:
In trouble, what a resting place
Have they who know the throne of grace!

John Newton: From 'Hannah' which begins 'When Hannah press'd with grief

The prophet's presence cheer'd their toil,
They watch'd the words he spoke;
Whether they turn'd the furrow'd soil,
Or fell'd the spreading oak.

John Newton: From his hymn on Elisha 'The borrowed Axe' which begins 'The prophets' sons, in time of old

The lion that on Samson roar'd
And thirsted for his blood;
With honey afterwards was stor'd
And furnish'd him with food.

John Newton: Opening verse of 'Samson's Lion

This selection of texts comes from the first book of *Olney Hymns*

In 1771, shortly after the death of his only son, William Bull came as Independent Minister to nearby Newport Pagnell: and a deep friendship began which lasted until Newton's death. William Bull also undertook to care for William Cowper when the Newtons finally left Olney. Newton records after meeting Bull: *I am struck with the wisdom, grace, and impression of Thine image which Thou hast given to Thy servant Bull, and I hope Thou wilt teach me to profit thereby. Surely I love him for Thy sake.*

. . . London is such a noisy, hurrying place, I wish you would leave it, fill your coach with those whom you love best, and come and spend a few days with us.

John Newton: An invitation from Olney to Daniel West and his family in a letter of 28th July 1772

Newton first came to know the Barhams in 1773. After staying at their home for a few days, he described them as *such a happy family perhaps I never saw, where the peace and love of God seem to dwell in every heart.* The name Barham had been added to John Foster – he was the son of a distinguished Northumbrian family of Fosters – when at an early age he became united to the Moravian Church. Newton found him, next to Bull, one of those most on the same spiritual wavelength as himself, and was glad he came to live about twelve miles distant.

Whoever is truly humbled will not be easily angry, will not be positive and rash, will be compassionate and tender to the infirmities of his fellow-sinners, knowing, that if there be a difference, it is grace that has made it, and that he has the seeds of every evil in his own heart; and, under all trials and afflictions, he will look to the hand of the Lord, and lay his mouth in the dust, acknowledging that he suffers much less than his iniquities have deserved.

John Newton: From 'Letters to a Nobleman' April 1772

A lion visited Olney, which seemed very tame. *I went to see him. He was wonderfully tame, as familiar and docile and obedient as a spaniel.* The keeper however told Newton the animal had its surly moments. The hymnwriter incorporated the incident into his hymn about Samson's lion; he wrote to his wife the lion's character was like his own, sweet and docile but with surly moments.

On Thursday 14th May 1773, Newton recorded in his diary *We are now free from company for the first time since January 20th.*

In 1774, Newton was in touch with Mrs. Talbot, widow of the Reverend W. Talbot, an evangelical minister of St. Giles, Reading. Apart from her grief over the loss of her husband, she was concerned that the theological views of his successor, the Hon. W. B. Cagogan were very different. Over the years the new appointee was to change his views considerably, perhaps in part due to the persistence of Mrs. Talbot.

I hope you are both well reconciled to the death of your child. Indeed, I cannot be sorry for the death *infants.* How many storms do they escape! Nor can I doubt, in my private judgement, that they are included in the election of grace.

John Newton in a letter of 22nd October 1773

John and Mary had been married nearly twenty five years when they became foster-parents. Mary's brother George, who had lost his wife when their baby was born, himself died: and Betsy Catlett came as a child to the Vicarage. Mary's ailing father needed care, too, and he spent the last eighteen months of his life in the care of his daughter and her husband at Olney. In 1775, England was hit by

. . . Peter little thought himself entitled to that supreme prerogative, as the immediate Vicar of Jesus Christ, which his pretended successors falsely ascribe to him; nor did his brethren remind him of his privilege, otherwise there could have been no debate, for his declaration would have been decisive; but, waiving the claim of authority, he argued the insignificance of the Jewish rites as to salvation, from the Lord's conduct towards Cornelius and his friends, by his ministry.
John Newton: From 'A Review of Ecclesiastical History'

earthquake and the hymn *Altho' on massy pillars built* records how the earth *trembled under Britain's guilt.*

Altho' on massy pillars built,
The earth has lately shook;
It trembled under Britain's guilt,
Before its Maker's look.

. . . . .

Repent before the Judge draws nigh;
Or else when he comes down,
Thou wilt in vain for earthquakes cry,
To hide thee from his frown.

John Newton: The first and fifth verses of 'On the Earthquake, September 8, 1775'

A happy incident that year was the arrival in Olney of Mr. Whitford from his Liverpool days, now an Independent minister, the visit reviving the *remembrance of many incidents long since past . . .* A little later, Whitford was appointed to the Independent Church at Olney. It was common practice in those days to preach sermons to the young at the beginning of the year, and at Olney Newton and his dissenting brethren did so on successive evenings. Newton records being impressed with Whitford's preaching.

You cannot, I trust, in conscience think of laying out one penny more than is barely decent; unless you have another penny to help the poor.
John Newton: From 'On Trust in the Providence of God, and Benevolence to his Poor'

The pulpit at Olney was filled quite often by the Reverend Matthew Powley, another colleague of John Newton in charge of the church at Staithwaite near Huddersfield. His visits to Olney were linked to his wife visiting her mother: he had married Mary Unwin's daughter, Susanna, in 1774.

I am a strange refractory patient; have too often neglected his prescriptions, and broken the regimen he appoints me to observe.
John Newton, concerning himself as a patient, in a letter of 2nd June 1772 to Daniel West

In 1776, Newton developed a tumour on his thigh and in consequence needed a surgical operation for its removal performed at Guy's Hospital in London. He wrote of this that he considered *the ability to bear a very sharp operation with tolerable calmness and confidence was a greater favour from God than deliverance from his malady.* That same year Thomas Beddoes discovered nitrous oxide which was eventually to mark the beginning of a new era.

That same year, an eclipse of the moon led to a hymn with a verse *How punctually eclipses move.* He wrote to the Reverend William Rose, later Rector of Beckenham in Kent and Vicar of Codrington *such a scene of prosperity as seems to lie before you, is full of snares, and calls for a double effort of watchfulness and prayer.* The same month, with

Mary Newton ill, he wrote in his diary: *We are still in the furnace, but I hope Thou, my gracious Lord, art near as a refiner of silver.*

How punctually eclipses move,
Obedient to thy will!
Thus shall thy faithfulness and love,
Thy promises fulfill.

Dark, like the moon without the sun,
I mourn thine absence, Lord!
For light or comfort I have none,
But what thy beams afford.

But lo! the hour draws near apace,
When changes shall be o'er;
Then I shall see thee face to face,
And be eclips'd no more.

John Newton: From 'On the Eclipse of the Moon. July 30, 1776' which begins 'The moon in silver glory shone'

In the nearby village of Weston Underwood, the curate, Reverend Thomas Scott came under Newton's influence. He was particularly impressed with Newton's care for all his parishioners: Newton's sympathy with the Methodist movement led Scott to confess the error of his ways in rejecting their ideas out of hand. He was eventually to become a leading evangelical, publishing a Bible commentary greatly respected in these circles and the controversial book *The Force of Truth* with which the hymnwriter Reginald Heber took issue. In 1777, a great fire struck Olney adding to the distress of Olney's poor, already greatly impoverished by the near collapse from hand lace-making as the result of machinery taking over.

In this Narrative, little more is contained than a history of the workings of my heart, that forge of iniquity; and of my conscience, that friendly monitor, whom we generally hate, because, as far as informed, it boldly tells us the truth, whom we endeavour to pacify, to lay asleep, and to render insensible, as if seared with a hot iron; which, through the deceitfulness of our hearts, of sin, and of the world, by the assistance of Satan, we generally in time accomplish, and to whose remonstrances, until this is effected, we commonly deafen ourselves, by living in a continual noise and bustle.

Thomas Scott: From his preface to 'The Force of Truth'

All this time, hymn after hymn was being written so that there was much organisation to be undertaken to prepare *Olney Hymns* for publication. A Church of England hymnbook was in those days a most unusual occurrence. It was expected by the Establishment that singing would be limited to metrical psalms. The first section is arranged in Biblical order, with the first eighty texts taking us through the Old Testament:

Genesis	12	Psalms	7
Exodus	6	Proverbs	2
Leviticus	1	Ecclesiastes	2
Numbers	1	Solomon's Song	1
Joshua	1	Isaiah	8
Judges	3	Jeremiah	3
I Samuel	4	Lamentations	1
II Samuel	3	Ezekiel	3
I Kings	6	Daniel	2
II Kings	4	Jonah	1
I Chronicles	1	Zechariah	4
Nehemiah	1	Malachi	1
Job	2		

For months and years of safety past,
Ungrateful, we, alas! have been;
Tho' patient long, he spoke at last,
And did the fire rebuke our sin.

The shout of fire! a dreadful cry,
Impress each heart with deep dismay;
While the fierce blaze and red'ning sky,
Make midnight wear the face of day.

The throng and terror who can speak?
The various sounds that fill'd the air!
The infant's wail, the mother's shriek,
The voice of blasphemy and pray'r!

But pray'r prevail'd, and sav'd the town;
The few, who lov'd the Saviour's name,
Were hear'd, and mercy hasted down
To change the wind, and stop the flame.

John Newton: From 'On the Fire at Olney. September 22, 1777' which begins 'Wearied by day with toils and cares'

Glorious things of thee are spoken,
Zion, city of our God!
He, whose word cannot be broken,
Form'd thee for his own abode:
On the rock of ages founded,
What can shake thy sure repose?
With salvation's walls surrounded
Thou mays't smile at all thy foes.

See! the streams of living waters
Springing from eternal love;
Well supply thy sons and daughters,
And all fear of want remove:
Who can faint while such a river
Ever flows their thirst t'assuage?
Grace, which like the Lord, the giver,
Never fails from age to age.

Round each habitation hov'ring
See the cloud and fire appear!
For a glory and a cov'ring,
Shewing that the Lord is near:
Thus deriving from their banner
Light by night and shade by day;
Safe they feed upon the Manna
Which he gives them when they pray.

Blest inhabitants of Zion,
Wash'd in the Redeemer's blood!
Jesus, whom their souls rely on,
Make them kings and priests to God:
'Tis his love his people raises
Over self to reign as kings
And as priests, his solemn praises
Each for a thank-off'ring brings.

Saviour, if of Zion's city
I thro' grace a member am;
Let the world deride or pity,
I will glory in thy name:
Fading is the worldling's pleasure,
All his boasted pomp and show;
Solid joys and lasting treasure,
None but Zion's children know.

*John Newton: 'Zion'
Tunes include ABBOT'S LEIGH and AUSTRIAN HYMN (AUSTRIA)*

This section contains Newton's well-known *Glorious things of thee are spoken*, with perhaps the rhyming of *rely on* with *Zion* showing the pressure of turning out at least one hymn in most weeks.

n the same section we find *One there is, above all others,* sparked off by Proverbs 28, 24 and *How sweet the name of Jesus sounds* which relates the name of Jesus to verse 3 of he first chapter of Solomon's Song.

One there is, above all others,
Well deserves the name of friend;
His is love beyond a brother's,
Costly, free, and knows no end:
They who once his kindness prove,
Find it everlasting love!

Which of all our friends to save us,
Could or would have shed their blood?
But our Jesus dy'd to have us
Reconcil'd, in him to God:
This was boundless love indeed!
Jesus is a friend in need.

Men, when rais'd to lofty stations,
Often know their friends no more;
Slight and scorn their poor relations
Tho' they valu'd them before.
But our Saviour always owns
Those whom he redeem'd with groans.

When he liv'd on earth abased,
Friend of sinners was his name;
Now, above all glory raised,
He rejoices in the same:
Still he calls them brethren, friends,
And to all their wants attends.

Could we bear from one another,
What he daily bears from us?
Yet this glorious Friend and Brother,
Loves us tho' we treat him thus:
Tho' for good we render ill,
He accounts us brethren still.

Oh! for grace our hearts to soften!
Teach us, Lord, at length to love;
We, alas! forget too often,
What a Friend we have above:
But when home our souls are brought,
We will love thee as we ought.

John Newton: 'A Friend that sticketh closer than a Brother'
Tunes include ALL SAINTS (German) and ST. LEONARD (Bach)

How sweet the name of Jesus sounds
In a believer's ear?
It soothes his sorrows, heals his wounds,
And drives away his fear.

It makes the wounded spirit whole,
And calms the troubled breast;
'Tis Manna to the hungry soul,
And to the weary rest.

Dear name! the rock on which I build,
My shield and hiding place;
My never-failing treas'ry fill'd
With boundless stores of grace.

By thee my pray'rs acceptance gain,
Altho' with sin defil'd;
Satan accuses me in vain,
And I am own'd a child.

Jesus! my Shepherd, Husband, Friend,
My Prophet, Priest, and King;
My Lord, my Life, my Way, my End,
Accept the praise I bring.

Weak is the effort of my heart,
And cold my warmest thought;
But when I see thee as thou art,
I'll praise thee as I ought.

'Till then I would thy love proclaim
With ev'ry fleeting breath;
And may the music of thy name
Refresh my soul in death.

John Newton: 'The Name of Jesus'
Tunes include MINSTER; RACHEL; ST. BERNARD (Tochter Sion, adapted Richardson); ST. BOTOLPH; ST. PETER; STRACATHRO)

I beg you to pray for me; I am a poor creature, full of wants. I seem to need the wisdom of Solomon, the meekness of Moses, and the zeal of Paul to enable me to make full proof of my ministry. But alas! you may guess the rest. Send me *The Way to Christ.*
John Newton writing to William Bull in April 1778

The sixty-one hymns linked to New Testament themes are distributed like this:

Matthew	11	I Corinthians	1
Mark	6	II Corinthians	1
Luke	13	Galatians	1
John	9	Philippians	1
Acts	6	Hebrews	4
Romans	2	Revelation	6

I heard him, and admired, for he could bring
From his soft harp such strains as angels sing, –
Could tell of free salvation, grace, and love,
Till angels listened from their home above.
John Newton: From 'The Olney Hymns' concerning his co-contributor, William Cowper

The second section on occasional subjects is divided under four broad themes covering *Seasons, Ordinances, Providences and Creation.* Whilst there are a few texts of the forty-two under *Seasons* which deal with the natural world, much space is devoted to hymns for use before and after annual sermons to young people on new-years' evenings. Many of Newton's texts again take Old Testament characters as starting points: like prophets of long-ago days, Newton was convinced that wars were a visitation from God. Thus, *Hark! how time's wide sounding bell* was written in 1778 against the background of the conflict between France and Britain. There is one text still sung which relates to the new life of a new season. *Pleasing Spring again is here*, with the first word usually changed to *Kindly* is sung to DA CHRISTUS GEBOREN WAR and ORIENTUS PARTIBUS (ST. MARTIN (Old French Melody)).

When Christ the *Corner-stone* stirreth himself in the *extinguished* Image of Man, in his hearty *Conversion* and *Repentance*, then Virgin *Sophia* appeareth in the stirring of the Spirit of Christ in the *extinguished Image*, in her Virgin's *Attire* before the Soul; at which the soul is so amazed and astonished in its *Uncleanness*, that all its *Sins* immediately awake in it, and it *trembleth* before her; for then the *Judgement* passeth upon the *Sins* of the Soul, so that it even goeth back on its Unworthiness, being *ashamed* in the Presence of its *fair Love*, and entereth into *itself, feeling* and *acknowledging* itself utterly *unworthy* to receiver such a *Jewel.*
From the translation of Jacob Boeheme's 'The Way to Christ': 'The Gates of the Paradisical Garden of Roses'

Twenty-one hymns under the theme *Ordinances*, used in the sense of a religious ceremonial observed, include seven for the Holy Communion, two on Prayer and two on Scripture. The best known text is Cowper's *Jesus, where'er thy people meet* written for the opening of a Prayer Room at The Great House, and Newton's *Chief Shepherd of Thy chosen sheep.* This house belonged to Lord Dartmouth but was unoccupied and several of its large rooms were make available for Church use.

Sixteen hymns in the section *Providences* indicate its use in the special sense of an act of divine intervention. Newton sees this in the commencement of the American War of Independence and the lining up of enemies against Britain, and in the recent experiences of fire and earthquake. This is the place for eight funeral hymns and two for meetings of Christian friends including *Kindred in Christ, for his dear sake.*

Verses selected from John Newton's Hymns based on St. Luke's Gospel.

How oft are we like Martha vex'd,
Encumber'd, hurried, and perplex'd?
While trifles to engross our thought,
The one thing needful is forgot.

From 'Martha and Mary' which begins 'Martha her love and joy express'd'

Wretches, who cleave to earthly things,
But are not rich to God;
Their dying hour is full of stings,
And hell their dark abode.

From 'The Worldling' which begins 'My barns are full, my stores increase'

Wonder and joy at once
Were painted in his face;
Does he my name pronounce?
And does he know my case?
Will Jesus deign with me to dine?
Lord, I, with all I have, am thine!

From 'Zaccheus' which begins 'Zaccheus climb'd the tree'

Yet I must blame while I approve,
Where is thy first, thy fervent love?
Does thou forget my love to thee,
That thine is grown so faint to me?

John Newton: From 'Ephesus' which begins 'Thus saith the Lord to Ephesus'

I know thy works, and I approve,
Tho' small thy strength, sincere thy love;
Go on, my word and name to own,
For none shall rob thee of thy crown.

John Newton: From 'Philadelphia' which begins 'Thus saith the holy One, and true'

To some he speaks as once of old,
I know thee, thy profession's vain;
Since thou art neither hot nor cold
I'll spit thee from me with disdain.

Thou boasteth, *I am wise and rich,*
Encreas'd in goods and nothing need;
And dost not know thou art a wretch,
Naked and poor, and blind and dead.

John Newton: From 'Laodicea' which begins 'Hear what the Lord, the great Amen'

As to evil thoughts, they as unavoidably arise from an evil nature as steam from a boiling tea-kettle . . .

John Newton: From a letter of 13th November 1772

By faith he is enabled to use prosperity with moderation . . .

John Newton: From 'Of the Practical Influence of Faith'

. . . the church of Rome, not merely by adopting an unmeaning burdensome train of ceremonies, but by her doctrines of papal infallibility, invocation of saints and angels, purgatory, absolution, the mass, and others of the like stamp, is become so exceedingly adulterated, that possibly some persons who may read these letters will form an unfavourable opinion of me, for declaring that I have not the least doubt but the Lord Jesus has had, from age to age, a succession of chosen and faithful witnesses within the pale of that corrupt church.

John Newton: From 'Aplogia: Letter II of Four letters to a Minister of an Independent Church by A Minister of the Church of England'

I have heard of a minister who used to compose hymns in the pulpit. It was his custom to give out one line; and by the time the congregation had sung the first, he had a second ready for them, and so on, as long as he thought proper to sing. These were not forms; they were composed ***pro re nata.*** **Before he had finished a second stanza, the former (as to the verse and cadence) was in a manner forgotten, and the same hymn was never heard twice. I know not what these unpremeditated pieces were in point of composition; but were I persuaded of the unlawfulness of forms of prayer, and at the same time approved the practice of singing in public worship, I should extremely covet the talent of extempore hymn-making, as one of the most necessary gifts a minister could possess, in order to maintain a consistency in his whole service.**

John Newton: From 'Apologia: Letter I of Four Letters to a Minister of an Independent Church by A Minister of the Church of England'

. . . when the love of Jesus is the constraining motive of our conduct, the necessary business of every day, in the house, the shop, or the field is ennobled, and makes a part of our religious worship . . .

John Newton: From a letter written in June 1776

Pleasing spring again is here!
Trees and fields in bloom appear;
Hark! the birds, with artless lays,
Warble their Creator's praise!
Where, in winter, all was snow,
Now the flow'rs in clusters grow;
And the corn, in green array,
Promises a harvest-day.

What a change has taken place!
Emblem of the spring of grace;
How the soul, in winter, mourns
Till the Lord, the Sun, returns;
Till the Spirit's gentle rain,
Bids the heart revive again;
Then the stone is turn'd to flesh,
And each grace springs forth afresh.

Lord, afford a spring to me!
Let me feel like what I see;
Ah! my winter has been long,
Chill'd my hopes, and stopp'd my song!
Winter threat'ned to destroy
Faith, and love, and ev'ry joy;
If thy life was in the root,
Still I could not yield thee fruit.

Speak, and by the gracious voice
Make my drooping soul rejoice;
O beloved Saviour, haste,
Tell me all the storms are past:
On thy garden deign to smile,
Raise the plants, enrich the soil;
Soon thy presence will restore
Life, to what seem'd dead before.

Lord, I long to be at home,
Where these changes never come!
Where the saints no winter fear,
Where 'tis spring throughout the year:
How unlike this state below!
There the flow'rs unwith'ring blow;
There no chilling blasts annoy,
All is love, and bloom, and joy.

John Newton: Spring II

Toplady: Who could bear to see that sight if there were not to be some compensation for these poor suffering animals in a future state?

Bull: I certainly hope that all the Bulls will go to heaven; but do you think this will be the case with all the animal creation?

Toplady: Yes, certainly, all, all.

Newton: What! do you suppose, sir, there will be fleas in heaven? for I have a special aversion to them.

A conversation between hymnwriter Augustus Toplady, author of 'Rock of Ages, cleft for me', William Bull and John Newton, in an inn at Olney, arising from bear-baiting outside the window. The incident occurred shortly before Toplady's death.

But conversation with most Christians is something like going to Court, where, except you are dressed exactly according to a prescribed standard, you will either not be admitted, or must expect to be heartily stared at; but you and I can meet and converse *sans constrainte*, in our undress, without fear of offending, or being counted offenders for a word out of place, and not exactly in the pink of the mode . . . I know not how it is; I think my sentiments and experience are as orthodox and Calvinistical as need be, and yet I am a sort of *speckled bird* amongst my Calvinist brethren.

John Newton writing to William Bull in 1778

Kindred in Christ, for his dear sake,
A hearty welcome here receive;
May we together now partake
The joys which only he can give!

To you and us by grace 'tis giv'n
To know the Saviour's precious name;
And shortly we shall meet in heav'n,
Our hope, our way, our end, the same.

John Newton: Opening verses of 'A Welcome to Christian Friends'

One speckled bird to another speckled bird, whom he loves most dearly, sendeth greeting.

William Bull begins his reply to John Newton's recent letter.

See how war, with dreadful stride,
Marches at the Lord's command;
Spreading desolation wide,
Thro' a once much-favour'd land:
War, with heart and arms of steel,
Preys on thousands at a meal;
Daily drinking human gore,
Still he thirsts, and calls for more.

John Newton: From 'Death and War, 1778' which begins 'Hark! how time's wide sounding bell'

When you are with the King, and getting good for yourself, speak a word for me and mine. I have reason to think you see Him oftener, and have nearer access to Him than myself. Indeed, I am unworthy to look at Him, or speak to Him at all, much more than He should speak tenderly to me. Yet I am wholly without His notice. He supplies all my wants, and I live under His protection. My enemies are His Royal Arms over my door, and dare not enter. Were I detached from Him for a moment, in that moment they would make an end of me.

John Newton writing to William Bull on 4th August 1778.

The book of nature open lies,
With much instruction stor'd;
But till the Lord anoints our eyes
We cannot read a word.

John Newton: Opening verse of 'The Book of Creation'

Sinner, art thou still secure?
Wilt thou still refuse to pray?
Can thy heart or hands endure
In the Lord's avenging day?
See, his mighty arm is bar'd!
Awful terrors clothe his brow!
For his judgment stand prepar'd
Thou must either break or bow.

John Newton: The opening verse of 'Prepare to meet God'

All twenty-one hymns under the heading *Creation* come from Newton's pen including *The book of nature open lies.* His texts recall the need for philosophers to have the Lord anoint their eyes if they are to view the universe with understanding. They draw inspiration from the rainbow, thunder, lightning, moon-light, the sea, flooding, thaw after snow, the spider, the bee, the tamed lion, sheep, a garden, the power to dream and the lodestone.

The third section on the rise, progress, changes and comforts of the spiritual life begins with five solemn addresses to sinners, all by Newton, including *Sinner, art thou still secure.* Next are nine texts, seeking, pleading, hoping, of which Newton wrote six. One of these, *Approach, my soul, the mercy-seat* is still sung sometimes to the tune BANGOR.

There follow twenty-eight texts under the general theme of *Conflict*, beginning with William Cowper's *Light shining out of Darkness* and its opening line *God moves in a mysterious way* born of the experience of despair which most who have experienced mental breakdown will have suffered. The best known of Newton's eighteen texts is *Begone unbelief* printed earlier in this chapter. Newton sometimes uses as an opening verse of a hymn an incident of the everyday world and follows it with a comparison in religious terms. Examples include *'Tis past – the dreadful stormy night* and *When the poor pris'ner thro' a gate.*

Approach, my soul, the mercy-seat
Where Jesus answers pray'r;
There humbly fall before his feet,
For none can perish there.

Thy promise is my only plea,
With this I venture nigh;
Thou callest burden'd souls to thee,
And such, O Lord, am I.

Bow'd down beneath a load of sin,
By Satan sorely prest;
By war without, and fears within,
I come to thee for rest.

Cowper and Newton each contributed eight hymns under the theme *Comfort.* The best known is Cowper's *Sometimes a light surprises.* Newton echoes a text in Deutero-Isaiah in the fourth verse of a hymn entitled *Confidence.* Its opening line is *Yes! since God himself has said it.* One of Newton's hymns under the heading *Dedication and Surrender* is still sung occasionally to SUNRISE. The scriptural reference Psalm 131, 2 and Matthew 18, 3 - 4 are cited against the title *The Child.* Its opening line *Quiet, Lord, my froward heart* uses a word for *perverse* no longer in current parlance. There follow twelve *Cautions* of which Newton's five include *The wishes that the sluggard frames*, for which Newton cites four verses from Proverbs (6, 10; 20, 4; 22, 13; 24, 30) as well as Luke 13, 24 and I Corinthians 9, 24. There are nine hymns of *Praise* of which Newton's six include *Rejoice, believer in the Lord* and *Let us love, and sing, and wonder* sung sometimes to-day to ABBEY and ALL SAINTS (German) respectively. The third section ends with seven short hymns to be used before the sermon, eight brief texts to be used after this discourse and five variants of the *Gloria Patri.* Of these short hymns, *May the grace of Christ our Saviour* is still widely used. Tunes include GOTT DES HIMMELS (WALTHAM), LANGDALE (Redhead); ST. CHRYSOSTOM, ST. OSWALD (ST. AMBROSE (Dykes)) and SHARON (Boyce) (BOYCE; HALTON HOLGATE). It is an effective memorial to the event when Newton was snatched by grace during the great storm off Ireland.

In due time, the Newtons with some reluctance sent Betsy to a boarding school at Northampton to which her uncle made regular visits: John was always good with children perhaps wanting to avoid the mistake of his own father who had found such difficulty in communicating with John. Later Betsy moved on to another school at Hampstead.

The Olney curate was not popular with everybody. He upset many by showing pro-American sympathies as the American War of Independence raged. He opposed the number of alehouses in the town. On Bonfire Night, a man would break the window of anyone in the town not putting a candle to shine from it, and John and Mary had to buy off the crowd threatening the vicarage. His congregation became smaller and

Be thou my shield and hiding-place!
That, shelter'd near thy side,
I may my fierce accuser face,
And tell him, *Thou hast dy'd.*

Oh wond'rous love! to bleed and die,
To bear the cross and shame;
That guilty sinners, such as I,
Might plead thy gracious name.

Poor tempest-tossed soul, be still,
My promis'd grace receive;
'Tis Jesus speaks – I must, I will,
I can, I do believe.

John Newton: 'The Effort – in another Measure'

Ah! Lord, since thou didst hide thy face,
What has my soul endur'd?
But now 'tis past, I feel thy grace,
And all my wounds are cur'd!

Oh wond'rous change! but just before
Despair beset me round;
I heard the lion's horrid roar,
And trembled at the sound.

John Newton: Verses from 'The Storm hushed' which begins ''Tis past – the dreadful stormy night'

As to all the doubts and questions,
Which my spirit often grieve,
There are Satan's sly suggestions,
And I need no answer give;
He would fain destroy my hope,
But the promise bears it up.

John Newton: From 'Confidence' which begins 'Yes! since God himself has said it'

Quiet, Lord, my froward heart,
Make me teachable and mild,
Upright, simple, free from art,
Make me as a weaned child:
From distrust and envy free,
Pleas'd with all that pleases thee.

John Newton: The opening verse of 'The Child(e)'

. . . I use you no worse than I am constrained to use many others whom I have long and dearly loved, and who have equal reason to say I am become a poor correspondent.

John Newton: From a letter of 12th June 1779

smaller, and when in 1779, he hinted he might leave Olney, there was no opposition to the idea! It was his wish that Thomas Scott should succeed him but this was rejected at first by the local people. However, they were not happy either with Newton's actual successor and when he resigned, Thomas Scott was appointed after all.

No hardship, he, or toil, can bear,
No difficulty meet;
He wastes his hours at home, for fear
Of lions in the street.

What wonder then if sloth and sleep,
Distress and famine bring!
Can he in harvest hope to reap,
Who will not sow in spring?

'Tis often thus, in soul concerns,
We gospel-sluggards see;
Who if a wish would serve their turns,
Might true believers be.

John Newton: Selected verses from 'The Sluggard' which begins 'The wishes that the sluggard frames'

If opposition has hurt many, popularity has wounded more.

John Newton: From 'On the Snares and Difficulties attending the Ministry of the Gospel'

Let us *love,* and *sing,* and *wonder,*
Let us *praise* the Saviour's name!
He has hush'd the Law's loud thunder,
He has quench'd mount Sinai's flame:
He has wash'd us with his blood,
He has brought us nigh to God.

Let us *love* the Lord who bought us,
Pity'd us when enemies;
Call'd us by his grace, and taught us,
Gave us ears, and gave us eyes:
He has wash'd us with his blood,
He presents our souls to God.

Let us *sing* tho' fierce temptations
Threaten hard to bear us down!
For the Lord, our strong salvation,
Holds in view the conqu'rors crown:
He who wash'd us with his blood,
Soon will bring us home to God.

Let us *wonder*, grace and justice,
Join and point to mercy's store;
When thro' grace in Christ our trust is,
Justice smiles, and asks no more:
He who wash'd us with his blood,
Has secur'd our way to God.

Let us *praise*, and join the chorus
Of the saints, enthron'd on high;
Here they trusted him before us,
Now their praises fill the sky:
Thou hast wash'd us with thy blood,
Thou art worthy, Lamb of God!

Hark! the name of Jesus, sounded
Loud, from golden harps above!
Lord, we blush, and are confounded,
Faint our praises, cold our love!
Wash our souls and songs with blood,
For by thee we come to God.

John Newton: 'Praise for redeeming Love'

Rejoice, believer, in the Lord
Who makes your cause his own;
The hope that's built upon his word,
Can ne'er be overthrown.

Tho' many foes beset your road,
And feeble is your arm;
Your life is hid with Christ in God,
Beyond the reach of harm.

Weak as you are, you shall not faint,
Or fainting shall not die;
Jesus, the strength of ev'ry saint
Will aid you from on high.

Tho' sometimes unperceiv'd by sense,
Faith sees him always near;
A Guide, a Glory, a Defence,
Then what have you to fear?

As surely as he overcame,
And triumph'd once for you;
So surely you, that love his name,
Shall triumph in him too.

John Newton: 'Perseverance'

May the grace of Christ our Saviour
And the Father's boundless love,
With the holy Spirit's favour,
Rest upon us from above!
Thus may we abide in union
With each other, and the Lord;
And possess, in sweet communion,
Joys which earth cannot afford.

John Newton

The church of St. Mary Woolnoth of the Nativity, to which John Newton was called was in the heart of the City of London. It occupies a site which has been a place of worship from the days of pre-history. When digging the foundations of the present building in 1716, evidence was found of a pagan fane, suggesting the site had once been the place for the rituals of nature-worshippers. Archaeologists unearthed more recently a tessellated pavement, dating from Roman times, and considered it to have been a remnant of a Temple of Concord. A timbered structure in Saxon times was replaced by William the Conqueror's Norman Church. In 1438 a new building was erected, being severely damaged in the Great Fire of London in 1666. Some repairs were done but during the reign of Queen Anne (1702 - 1714) the church was declared unsafe. The parishioners persuaded Parliament to use funds from an excise duty on coal imposed for rebuilding London after the Great Fire to finance the building of a new Church. The architect appointed was Nicholas Hawksmoor, perhaps Sir Christopher Wren's most

Can you provide a breakfast on Wednesday morning between seven and eight o'clock for two hungry parsons? Mr. Scott and I think to call upon you about that time, on our way to the visitation at Stratford. Perhaps you will get your gray horse ready from Bury Field, and take a ride with us.

John Newton requesting hospitality from his friend William Bull for Rev. Thomas Scott and himself, July 1779

How the world goes, I know not; for I seldom see a news-paper for a fortnight together; when I do, I meet with so little to please me, that I seem rather to prefer a state of ignorance, which gives me more scope for hoping for the best.

John Newton, in a letter of April 1780

brilliant pupil, and the foremost exponent of the English Baroque style. The church was completed in 1727 and consecrated to the glory of God on Easter Day.

John and Mary settled into a new home at Charles Square Hoxton. From London, John relayed news to William Cowper, with whom he had worked so closely on the *Olney Hymnbook*. In return Cowper kept the Newtons in touch with what went on in Olney. Fish was sent from London a hen came by way of a return gift: Newton's reputation and writings attracted a reasonably-sized congregation but there were not many from his own parish. Consequently, soon after he arrived, he circulated a pastoral letter to his parishioners: it bore the text from Acts 26, 3, *I beseech thee to hear me patiently.*

Charles Square was full of people on Monday the 5th; but they behaved peaceably, made a few inquiries and soon went away. The devastations on Tuesday and Wednesday nights were horrible. We could count from our back windows six or seven terrible fires each night, which, though at a distance, were very affecting. On Wednesday night and Thursday the military arrived, and saved the city, which otherwise, I think, would before this time have been in ashes from end to end.

John Newton writing to William Bull in the summer of 1780 about the No-Popery riots in London

The cloud of smoke hanging over London, to which every house contributed its quota, led me to moralise. I thought it an emblem of the accumulated stock of misery, arising from all the trials and afflictions of individuals within my view.

John Newton describes a trip to Greenwich on 30th June 1780

The only cause of complaint, or rather of grief, which you have given me is, that so many of those, to whom I earnestly desire to be useful, refuse me the pleasure of seeing them at church on the Lord's day. My concern does not arise from the want of hearers. If either a numerous auditory, or the respectable characters of many of the individuals who compose it, could satisfy me, I might be satisfied. But I must grieve, while I see so few of my own parishioners among them . . .

. . . Of late years the name of Methodist has been imposed as a mark and vehicle of reproach. I have not hitherto met with a person who could give a difinition or precise idea of what is generally intended by the formidable word, by those who use it to express their disapprobation. Till I do, I am at a loss whether to confess or deny that I am (what some account me) a methodist. If it be supposed to include any thing, whether in principle or conduct, unsuitable to the character of a regular minister of the church of England, I may, and I do, disown it. And yet is probable, that some of my parishioners hearing, and easily taking it for granted, that I am a methodist, think it sufficient proof that it cannot be worth their while to hear me.

That I may not disgust and weary my hearers by the length of my sermons, I carefully endeavour not to exceed three quarters of an hour, at those seasons when I have most reason to hope for the presence of my parishioners. At other times I allow myself a longer term; but even this, I understand, is thought too long.

From John Newton's pastoral letter 'A Token of Affection and Respect' to the Parishioners of St. Mary Woolnoth and St. Mary Woolchurch in the City of London, 1781

St. Mary Woolnoth, City of London

I pity such wise-headed Calvinists as you speak of. I am afraid there are no people more fully answer the character, and live in the spirit of the Pharisees of old, than some professed loud sticklers for free grace.

John Newton: From a letter of 29th December 1780

In responding to an enquiry about the state of religion in London at that time, he reported entirely from his evangelical viewpoint. In the Established Church, he had just one gospel minister colleague with a church of his own. (That was the Reverend Romaine of Blackfriars). There were however ten other gospel ministers around who preached from time to time in fifteen or sixteen churches. He mentioned the Tabernacle and Tottenham Court Chapel, both very large and in the hands of Mr. Whitefield's trustees, and in which the *Gospel is dispensed to many thousands of people, by a diversity of ministers, clergy, dissenters, or lay-preachers, who are, in general, lively, faithful, and acceptable men.* He also mentioned the Lock Chapel besides a second unnamed centre in Westminster and Lady Huntingdon's Chapel, this last seating two thousand. Of Mr. Wesley, he noted he had one large chapel, and several smaller; and, despite their being *Armenians, as we say, there are many excellent Christians, and some good preachers amongst them.* He added a reference to several other preachers who might be called independent methodists.

For, though my regard to the authority of the great Lord and Lawgiver of the church, did not directly oblige me to unite with the establishment, it discouraged me from uniting with any of the parties who pretended an exclusive right from Him to enforce their own particular church forms.

John Newton on reasons for his not being a dissenter in 'Apologia: Four letters to A Minister of an Independent Church by A Minister of the Church of England'

Most of the Presbyterian ministers failed, in Newton's view, to preach the doctrines of the Cross. The Baptists received his general approval but were over zealous on the point of baptism. The Independents, looked at as a whole, were a group amongst whom life and glory was abated. He ended his survey with a reference to *settlements of the Unitas Fratrum, the Brethren, or as they are more vulgarly called, the Moravians.* He wrote of them as persons against whom prejudice was popular and he acknowledged that at one time he shared it. Though he could not agree with them in all things, he was able to declare: *I do not know more excellent, spiritual, evangelical people in the Lord.*

. . . Their views are so strict, that if they certainly knew that a person who wished to communicate with them was the most eminent Christian in the land, unless he was likewise baptized in their manner, they could not, they durst not, admit him to the Lord's table, to eat of that bread, and to drink of that cup, which is, by his command and appointment, the privilege and portion of all believers.

John Newton about some Baptists in 'Apologia: Four letters to A Minister of an Independent Church by A Minister of the Church of England'

It will be remembered that John Newton was self-educated in those subjects which might have been the subject of a theological training at College. In one of his imaginary letters under the name of Omicron, he discusses the skills he thinks a tutor should give a prospective minister of the Church; and he warns tutors to protect students from the temptations of female entanglements. These letters were first published in the *Gospel Magazine* and later published as a collection.

. . . strive and pray against indolence, look upon it as a hurtful, yea, a sinful thing. Read in English and French, write and work. Your mamma and I will be both willing you should diversify these employments as may be most agreeable to your own inclination; but we would not wish to see you idle.

John Newton writing to Betsy on 10th January 1781

The study of the works of God, independent of his word, though dignified with the name of *philosophy*, is no better than an elaborate trifling and waste of time.

John Newton: From 'A Plan of a compendious Christian Library'

They have gone to the academy humble, peaceable, spiritual, and lively; but have come out self-wise, dogmatical, censorious, and full of a prudence founded upon the false maxims of the world.

John Newton: Extract of a letter to a Student in Divinity.

I do not envy you your pleasure with Dr. Kennicott. One hundred and fifty folio pages in Latin, and upon a critical subject, would have taken me a year instead of a month to wade through. I have lost my acumen for such disquisitions, and perhaps I am as well without it.

John Newton writing to William Bull, in February 1781 about Dr. Kennicott's 'Prolegomena'

And he cannot be too careful, both by advice and vigilance, to prevent them from forming any female connexions while under his roof, however honourable the views, or deserving the person, may be. Love and courtship are by no means favourable to study, nor indeed to devotion, at a time when their present engagements, and the uncertainty of their prospects in future life, render a settlement by marriage improper, if not impracticable . . . . . I am not the son of a prophet, nor was I bred up among the prophets. I am quite a stranger to what passes within the walls of colleges and academies. I was as one born out of due time, and led, under the secret guidance of the Lord, by very unusual steps, to preach the faith which I once laboured to destroy . . . . . For his first essential, indispensable qualification, I require a mind deeply penetrated with a sense of the grace, glory, and efficacy of the Gospel. However learned and able in other respects, he shall not have a single pupil from me, unless I have reason to believe that his heart is attached to the person of the Redeemer as God-man; that, as a sinner, his whole dependence is upon the Redeemer's work of love, his obedience unto death, his intercession and mediatorial fulness . . . . . Besides an accurate skill in the school classics, he should be well acquainted with books at large, and possessed of a general knowledge of the state of literature and religion, and the memorial events of history in the successive ages of mankind. Particularly, he should be well versed in ecclesiastical learning . . . . . Calvin, Turretin, Witsius, and Ridgeley, are those with whom I have formerly been most acquainted. But, indeed, of these, at present, I can remember little more than that I have read them, or the greatest part of them. I recollect just enough to say, that though I approve and admire them all, I have at the same time my particular objections to them all, as to this use of them. The Bible is my body of divinity; and were I a tutor myself, I believe I should prefer the Epistles of St. Paul, as a summary, to any human systems I have seen, especially his Epistles to the Romans, Galatians, the Hebrews and Timothy.

John Newton: From 'A Plan of Academical Preparation for the Ministry'

Newton's friend William Bull was a keen theologian: by now John Newton had lost the will to read theology in Latin. Dr. Kennicott was a leading Biblical scholar whose Bible annotated with copious notes came to be widely used. The Reverend Martin Madan, a cousin of William Cowper, who had shown great interest in Newton at the time the seafarer was seeking ordination, and who until now had earned the respect of Bull and Newton, shocked them both by his publication of *Thelyphtor*, a most controversial book on the rights of women and, in particular, the responsibilities of men who chose to have sex with any woman. The publication led both Newton and Bull to dissociate themselves from Madan.

It is impossible to believe that there is a God, unless we acknowledge that he made the world; therefore those who reject the first verse in the Bible ought not to read any further.

Benjamin Kennicott

In 1783 *The Eclectic Society* began regular meetings at the *Castle and Falcon.* Three of this small group were John Newton, the Reverend Richard Cecil and the Reverend Henry Foster. Later, they met at St. John's Chapel, Bedford Row. After tea and a short prayer, they spent three hours in conversation: Newton kept notes based on these for the next ten years.

. . . it appears that *marriage*, as instituted of God, simply consists (as to the essence of it) in the *union* of the *man* and *woman* as *one body*; for which plain and evident reason, no outward forms or ceremonies of man's invention, can add to or diminish from the effects of *this union* in the sight of God. What ends these things may serve, as to civil purposes, I shall not dispute: but I cannot suppose the *matrimonial*-service in our church, or any other, can make the parties more *one flesh* in the sight of God, supposing them to have been *united*, than the *burial*-service can make the corpse over which it is read more dead than it was before.

Martin Madan: From 'Thelyphtor'

It seems likely that on one of his visits from Olney to London to preach, his congregation had included a young lad by the name of William Wilberforce. Hearing about this, William's mother had quickly acted to remove her son from the influence of 'turning Methodist', a phrase which spilled over as an appropriate term of derision as applicable to all Evangelicals. John Pollock is his biography of Wilberforce wrote that *the fashionable world of 1785 looked at Newton and other Evangelicals with the contempt, suspicion and ignorance that Soviet Russia reserves for its Jewish and Christian believers.*

William Wilberforce, the young man having bought his way into Parliament in 1780, was by the Autumn of 1785, suffering an agony of conscience. If his friends praised his morality, he knew it was but skin-deep. *I was filled with sorrow. I am sure that no human creature could suffer more than I did for some months. It seems indeed it quite affected my reason: not so as others would observe, for all this time, I kept out of company. They might see I was out of spirits* . . . At the heart of Wilberforce's dilemma was that he had understood what the Methodists were proclaiming: if he became a Christian, he must put himself completely at God's disposal. What then would be his position with his friends? What of his political future? He corresponded about his dilemma with William Pitt: but felt the need for a confidant who would really understand his spiritual agony.

I long aimed to be something. I now wish I was more heartily willing to be nothing.

John Newton: From a letter of 13th March 1781

Cannot you contrive to put your lines in a little closer together? Your paper looks like a half-furnished room . . .

John Newton: From a letter to Betsy on 12th May 1783

I long to see you, and especially now, that we may read Mr. Gray's Elegy together . . .

John Newton writing to Betsy on 19th May 1783

When I mentioned my condescension to a child in creeping about the floor to please him, I thought you would probably call upon me for an explanation. You say directly after, you had been asked to go to the playhouse. But my compliance with the child must have some bounds. If he should choose to creep under the grate and burn his hair, I would prevent him if I could. But if that were out of my power, my regard for my own clothes would not permit me to be so complaisant as to follow him. If people will go to the play, let them honestly confess that their pleasure lies there, and not pretend a desire to become all things to all men with a view to do them good. The same argument might lead them to other places of bad resort. I do suppose it is possible for a person who has been used to the theatre in the time of ignorance to continue the custom some time after his being awakened, for want of reflection.

John Newton writing to Jane Dawson née Flower on 6th August 1783

The very act of consulting with the preacher he had heard as a boy was a decision of *ten thousand doubts*. He walked twice round Charles Square, Hoxton, before summoning the courage to actually knock at the Rector's door. Newton pointed the way ahead. He was to write nearly two years later: *It is hoped and believed that the Lord raised you up for the good of His Church and for the good of the nation.* Wilberforce had a hard winter but at Easter, visited the Reverend William Unwin, the rector of Stock, the son of Mary Unwin, with whom William Cowper spent so many years. Soon after the sun had risen on Easter Day in 1786, as Wilberforce wrote in a letter, he took to the fields to pray and give thanks *amidst the general chorus with which all nature seems on such a morning to be swelling the song of praise of thanksgiving.*

William Wilberforce was a frequent member of Newton's congregation at St. Mary Woolnoth of the Nativity: here the anti-slavery campaign was cradled. Newton's own experiences, to be printed in *Thoughts upon the African Slave Trade*, provided powerful ammunition to the cause. Nevertheless there were to be many setbacks and disappointments ahead.

However, my stay in Antigua and St. Christopher's (the only islands I visited) was too short, to qualify me for saying much, from my own certain knowledge, upon this painful subject. Nor is it needful: – enough has been offered by several respectable writers, who have had the opportunity of collecting surer and fuller information.

One thing I cannot omit, which was told me by the gentleman to whom my ship was consigned, at Antigua, in the year 1751, and who was himself a planter. He said, the calculations had been made, with all possible exactness, to determine which was the preferable, that is, the more saving method of managing slaves:

'Whether to appoint them moderate work, plenty of provision, and such treatment as might enable them to protract their lives to old age?'

Or,

'By rigorously straining their strength to the utmost, with little relaxation, hard fare, and hard usage, to wear them out before they became useless, and unable to do service; and then, to buy new ones, to fill up their places?'

He farther said, that these skilful calculators had determined in favour of the latter mode, as much the cheaper . . .

. . . . .

In the Portuguese ships, which trade from Brasil to the Gold coast and Angola, I believe, a heavy mortality is not frequent. The slaves have room, they are not put in irons (I speak from information only), and are humanely treated.
With our ships, the great object is, to be full. When the ship is there, it is thought desirable she should take as many as possible. The cargo of a vessel of a hundred tons, or little more, is calculated to purchase from two hundred and twenty to two hundred and fifty slaves. Their lodging-rooms below the deck, which are three (for the men, the boys, and the women), beside a place for the sick, are sometimes more than five feet high, and sometimes less; and this height is divided towards the middle, for the slaves lie in two rows, one above the other, on each side of the ship, close to each other, like books upon a shelf. I have known them so close, that the shelf would not, easily, contain one more. And I have known a white man sent down, among the men, to lay them in these rows to the greatest advantage, so that as little space as possible might be lost.
Let is be observed, that the poor creatures, thus cramped for want of room, are likewise in irons, for the most part both hands and feet, and two together, which makes it difficult for them to turn or move, to attempt either to rise or to lie down, without hurting themselves, or each other. Nor is the motion of the ship, especially her heeling, or stoop on one side, when under sail, to be omitted; for this, as they lie athwart, or cross the ship, adds to the uncomfortableness of their lodging, especially to those who lie on the leeward or leaning side of the vessel.

Thus I lived at Olney; how different is London! But hush! Olney was the place once, London is the place now. Hither the Lord brought me, and here he is pleased to support me, and in some measure (I trust) to own me.
John Newton: From a letter of May 1784

Dire is the tossing, deep the groans –

The heat and the smell of these rooms, when the weather will not admit of the slaves being brought upon deck, and of having their rooms cleaned every day, would be almost insupportable to a person not accustomed to them. If the slaves and their rooms can be constantly aired, and they are not detained too long on board, perhaps there are not many die; but the contrary is often their lot. They are kept down, by the weather, to breathe a hot and corrupted air, sometimes for a week: this, added to the galling of their irons, and the despondency which seizes their spirits when thus confined, soon becomes fatal. And every morning, perhaps, more instances than one are found, of the living and the dead, like the captives of Mezentius, fastened together.
Epidemical fevers and fluxes, which fill the ship with noisome and noxious effluvia, often break out, and infect the seamen likewise, and thus the oppressors, and the oppressed, fall by the same stroke. I believe, nearly one-half of the slaves on board, have, sometimes, died; and that the loss of a third part, in these

But where is heaven? Is it at an immense distance beyond the fixed stars? Have our ideas of space anything to do with it? Is not heaven often upon earth in proportion as the presence of God is felt?
John Newton: From a letter of 25th February 1785

circumstances, is not unusual. The ship, in which I was mate, left the coast with two hundred and eighteen slaves on board; and though we were not much affected by epidemical disorders, I find by my journal of that voyage (now before me), that we buried sixty-two on our passage to South Carolina, exclusive of those which died before we left the coast, of which I have no account.

I believe, upon an average between the more healthy, and the more sickly voyages, and including all contingencies, one fourth of the whole purchase may be allotted to the article of mortality: that is, if the English ships purchase *sixty thousand slaves* annually, upon the whole extent of the coast, the annual loss of lives cannot be much less than *fifteen thousand*.

I am now to speak of the survivors – When the ships make the land (usually the West India islands), and have their port in view, after having been four, five, six weeks, or a longer time, at sea (which depends much upon the time that passes before they can get into the permanent trade-winds, which blow from the north-east and east across the Atlantic), then, and not before, they venture to release the men slaves from their irons: and then, the sight of the land, and their freedom from long and painful confinement, usually excite in them a degree of alacrity, and a transient feeling of joy –

There is no new thing under the sun. When I read Sallust's account of the Jugurthine war, I seem to read (mutatis mutandis) our own history. The wealth and luxury which followed the successes of Lucellus in Asia soon destroyed all appearance of public spirit in Rome. Our acquisitions in the East have had a similar effect.
John Newton: 'A letter on Political Debate'

The prisoner leaps to lose his chains.

But this joy is short-lived indeed. The condition of the unhappy slaves is in a continual progress from bad to worse. Their case is truly pitiable, from the moment they are in a state of slavery in their own country; but it may be deemed a state of ease and liberty, compared with their situation on board our ships.

Yet, perhaps they would wish to spend the remainder of their days on ship-board, could they know, beforehand, the nature of the servitude which awaits them on shore; and that the dreadful hardships and sufferings they have already endured, would, to the most of them, only terminate in excessive toil, hunger, and the excruciating tortures of the cart-whip, inflicted at the caprice of an unfeeling overseer, proud of the power allowed him of punishing whom, and when, and how he pleases.

. . . . .

Did you possess the gift of foresight, and think to save your credit upon easy terms, when you promised to lodge in Charles Square next time you came to town, if we could receive you? I thought no more of removing into the city when I saw you last than of going to Bengal. But sure enough, I have taken a house in Coleman Street Buildings, and it will be mine at Lady-Day.
John Newton writing to 'Mon cher Taureau' (William Bull) about his move in 1786

Usually, about two-thirds of a cargo of slaves are males. When a hundred and fifty or two hundred stout men, torn from their native land, many of whom never saw the sea, much less a ship, till a short space before they are embarked; who have, probably, the same natural prejudice against a white man, as we have against a black; and who often bring with them an apprehension they

are bought to be eaten: I say, when thus circumstanced, it is not to be expected that they will tamely resign themselves to their situation. It is always taken for granted, that they will attempt to gain their liberty if possible. Accordingly, as we dare not trust them, we receive them on board, from the first, as enemies; and, before their number exceeds, perhaps, ten or fifteen, they are all put in irons; in most ships, two and two together. And frequently, they are not thus confined, as they might most conveniently stand or move, the right hand and foot of one to the left of the other, but across; that is, the hand and foot of each on the same side, whether right or left, are fettered together: so that they cannot move either hand or foot, but with great caution, and with perfect consent. Thus they must sit, walk, and lie, for many months (sometimes for nine or ten), without any mitigation or relief, unless they are sick.
In the night, they are confined below; in the daytime (if the weather be fine) they are upon deck; and as they are brought by pairs, a chain is put through a ring upon their irons, and this is likewise locked down to the ring-bolts, which are fastened, at certain intervals, upon the deck. These, and other precautions, are no more than necessary; especially, as while the number of slaves increases, that of the people who are to guard them, is diminished, by sickness, or death, or by being absent in the boats: so that, sometimes, not ten men can be mustered, to watch, night and day, over two hundred, besides having all the other business of the ship to attend.

***Moderation* is a Christian grace; it differs much from that tame, unfeeling neutrality between truth and error, which is so prevalent in the present day. As the different rays of light which, when separated by a prism, exhibit the various colours of the rainbow, form, in their combination, a perfect and resplendent *white*, in which every colour is incorporated; so, if the graces of the Holy Spirit were complete in us, the result of their combined effect would be a truly candid, moderate, and liberal spirit towards our brethren.**

John Newton: From the preface to 'Messiah: Fifty Expository Discourses, on the series of Scriptural Passages Which form the Subject of the celebrated Oratorio of Handel'

I hope it will always be a subject of humiliating reflection to me, that I was once an active instrument in a business at which my heart now shudders. My headstrong passions and follies plunged me, in early life, into a succession of difficulties and hardships, which, at length, reduced me to seek a refuge among the natives of Africa. There, for about the space of eighteen months, I was in effect, though without the name, a captive, and a slave myself; and was depressed to the lowest degree of human wretchedness. Possibly I should not have been so completely miserable, had I lived among the natives only, but it was my lot to reside with white men; for at that time several persons of my own colour and language were settled upon that part of the Windward coast which lies between Sierra Leon and Cape Mount; for the purpose of purchasing and collecting slaves, to sell to the vessels that arrived from Europe.
This is a bourn from which few travellers return, who have once determined to venture upon a temporary residence there; but the good providence of God, without my expectation, and almost against my will, delivered me from those scenes of wickedness and woe; and I arrived at Liverpool, in May 1748. I soon revisited the place of my captivity, as mate of a ship, and, in the year

It is not easy for those whose habits of life are insensibly formed by the customs of modern times, to conceive any adequate idea of the pastoral life, as it obtained in the eastern countries, before that simplicity of manners, which characterized the early ages, was corrupted by the artificial and false refinements of luxury.

John Newton: From 'Messiah' 'Sermon 13: The Great Shepherd', based on Isaiah 40, 11.

The Kingdom of our Lord in the heart, and in the world, is frequently compared to a building or house, of which he himself is both the foundation and the architect. A building advances by degrees, and while it is in an unfinished state, a stranger cannot, by viewing its present appearance, form an accurate judgement of the design, and what the whole will be when completed. For a time, the walls are of unequal height, it is disfigured by rubbish, which at the proper season will be taken away; and by scaffolding, which, though useful for carrying on the building, does not properly belong to it, but will likewise be removed when the present temporary service is answered. But the architect himself proceeds according to a determinate plan, and his idea of the whole work is perfect from the beginning. It is thus the Lord views his people in the present life.

John Newton: From 'Messiah': 'Sermon 37: The Extent of Messiah's Spiritual Kingdom' based on Revelation 11, 15

. . . the pointing of the New Testament, though it has a considerable influence upon the sense, is of inferior authority. It is a human invention, very helpful, and for the most part, I suppose, well executed. But in some places it may admit of real amendment.

John Newton: From 'Messiah' 'Sermon 46: 'Accusers Challenged' based on Romans 8, 33

1750, I was appointed commander; in which capacity I made three voyages to the Windward coast for slaves.
I first saw the coast of Guinea, in the year 1745, and took my last leave of it in 1754. It was not, intentionally, a farewell; but, through the mercy of God, it proved so. I fitted out for a fourth voyage, and was upon the point of sailing, when I was arrested by a sudden illness, and I resigned the ship to another captain. Thus I was unexpectedly freed from this disagreeable service. Disagreeable I had long found it; but I think I should have quitted it sooner, had I considered it as I now do, to be unlawful and wrong. But I never had a scruple upon this head at the time; nor was such a thought once suggested to me by any friend. What I did I did ignorantly; considering it as the line of life which Divine Providence had allotted me, and having no concern, in point of conscience, but to treat the slaves, while under my care, with as much humanity as a regard to my own safety would admit.

John Newton: 'Thoughts upon the African Slave Trade'

By our blood in Afric wasted,
Ere our necks receiv'd the chain;
By the mis'ries we have tasted,
Crossing in your barks the main;
By our suff'rings since ye brought us
To the man-degrading mart;
All sustain'd by patience, taught us
Only by a broken heart:

Deem our nation brutes no longer
Till some reason ye shall find
Worthier of regard and stronger
Than the colour of our kind.
Slaves of gold, whose sordid dealings
Tarnish all your boasted pow'rs,
Prove that you have human feelings,
Ere you proudly question ours!

William Cowper: From 'The Negro's Complaint'

Through the 1780's Handel's *Messiah* was popular with audiences in churches and concert halls alike. Newton chose to describe it as *of all our musical compositions, the most improper for a public entertainment*: he preached a year's sermons on the passages of scripture in the oratorio to redeem them to their proper use.

'Comfort Ye' from the 'Messiah' - Handel

. . . of all our musical compositions, this is the most improper for a public entertainment. But while it continues to be equally acceptable, whether performed in a church or in the theatre, and while the greater part of the performers and the audience are the same at both places, I can rate it no higher than one of the many fashionable amusements which mark the character of this age of dissipation. Though the subject be serious and solemn in the highest sense, yea, for that very reason, and though the music is, in a striking manner, adapted to the subject, yet, if the far greater part of the people who frequent the Oratorio, are evidently unaffected by the Redeemer's love, and uninfluenced by his commands, I am afraid it is no better than a profanation of the name and truths of God, a crucifying the Son of God afresh.

John Newton: From 'Messiah': 'Sermon 50: The Universal Chorus'

Though sinners are destitute of spiritual life, they are not therefore mere machines. They have a power to do many things, which they may be called upon to exert. They are capable of considering their ways; they know they are mortal; and the bulk of them are persuaded in their consciences, that after death there is an appointed judgement: they are not under an inevitable necessity of living in known and gross sins; that they do so, is not for want of power, but for want of will.

John Newton: From 'On the Propriety of a Ministerial Address to the Unconverted'

If her *(Eliza's)* recovery could be purchased, I think I would bid as high for it as my ability would reach, provided it was the Lord's will. But I am so short-sighted, that I dare not ask for the continuance of her life, (nor even of yours,) but with a reserve of submission to his wisdom.

John Newton, writing to his wife about Eliza Cunningham on 27th August 1785, while they were at Southampton.

John and Mary then suffered the loss of their neice Eliz Cunningham after a long illness and they found this a grea blow. At this time, the Newtons decided to move nearer t the Church, leaving Charles Square for Coleman Stree Buildings.

Dr. Ford ministered at Melton Mowbray for some twent years before losing his sense of zeal to the great concern of neighbour who alerted John Newton, who composed friendly but firm letter of reproof. Happily Dr. Ford's sens of ministry and mission was restored.

John Newton was one of eight pall bearers at the funeral o the great Methodist hymnwriter, Charles Wesley, like Newto ordained within the Church of England. Wesley ha nominated Newton for this task, which he carried throug though ill at the time and despite the bad weather at the tim of the burial.

John Newton was to have a considerable influence on th life stories of two other well-known figures. The first wa Claudius Buchanan. Nine years after leaving his family behin in Scotland, he again became interested in religion, and wa eventually directed to St. Mary Woolnoth. He foun Newton's sermons interesting, but on their own, he did no find the way ahead clear. He left a note for Newton which however, gave no address. Newton mentioned it in churc and asked the writer if present to visit him. In due tim Buchanan was converted and was ordained in 1795, travellin the following year as a chaplain to the East India Compan serving at Barrackpore. He was to serve as Vice-Provost o the College of Fort William also in Bengal. Some years later Buchanan was concerned about the Inquisition operated b the Roman Catholic Church in the province of Goa. Witl great courage he visited Josephus Doloribus in his ow palace and found several of his worst fears of torture an inhumane imprisonment well founded.

In the College of Fort-William in Bengal, there was a department for translating the Scriptures into the Oriental languages; and, so early as 1805 (the fifth year of its institution) a commencement had been made in five languages. The first version of any of the Gospels in the *Persian* and *Hindostanee* languages which were printed in India, issued from the Press of the College of Fort-William. The Persian was superintended by Lieut.-Colonel Colebrooke, and the Hindostanee by William Hunter, esq. The Gospels were translated into Western *Malay* by Thomas Jarrett, esq. of the Civil Service; into the *Orissa* language by Pooroosh Ram, the Orissa Pundit; and into the Mahratta language by Vydyunath, the Mahratta Pundit, under the superintendence of Dr. William Carey.

Claudius Buchanan: From 'Christian Research in Asia'

Hannah More was in one sense an unexpected person t become a friend of John Newton. Her earliest friends i London were David Garrick and his wife Eva Maria Veigel She mixed with the top literary set of the day: yet foun a leaning towards religion, becoming a writer of religiou tracts, poems and stories, and giving generously to the poor She was however unconcerned about the fairness of the grea divide between the haves and have-nots. Just as Newto earlier had been unconcerned by the existence of bondma and free-man, so Hannah More was unconcerned with th incidence of wealth and poverty. What was of over-ridin nportance was personal sin and the eed of redemption. She came to be nown as a light shining in a dark place, s many called the fashionable world. he was however concerned with the eachings of John Calvin which she ould neither accept nor find compatible vith Holy Scripture. She sought Newton's special help on this issue. Her health was not good and Newton, espite his years, records visiting her at Cowslip Green during one of her bouts f ill health.

Perish th'illiberal thought which wou'd debase
The native genius of the sable race!
Perish the proud philosophy, which fought
To rob them of the pow'rs of equal thought!
Does then th'immortal principle within
Change with the casual colour of a skin?
Does matter govern spirit? or is mind
Degraded by the form to which 'tis join'd?

No: they have heads to think, and hearts to feel,
And souls to act, with firm, tho' erring zeal;
For they have keen affections, kind desires,
Love strong as death, and active patriot fires;
All the rude energy, the fervid flame,
Of high-soul'd passion, and ingenuous flame:
Strong, but luxuriant virtues boldly shoot
From the wild vigour of a savage root.

Hannah More: From 'Slavery', 1788

One Sunday, Newton found an unusual equest for prayer. *A young man, having ome into the possession of a very onsiderable fortune, desires the prayers f the congregation, that he may be reserved from the snares to which it exposes him.* He was leased to commend the young man to God. On the other and, he was stopped on the steps of the church by a lady vho sought his congratulations as she had just won £2,500 in lottery. *Madam,* he said, *as a friend under temptation, I will ndeavour to pray for you.*

Whilst Newton had no temptation to turn politician, he was evertheless not angered in his latter years just by slavery ut by the laws which left over two hundred offences still unishable by death including stealing property worth more han five shillings. Many Bishops supported such laws: ot so this parish pastor.

. . . it is painful to see mothers, and possibly sometimes grandmothers, who seem, by the gaudiness and levity of their attire, very unwilling to be sensible that they are growing older.

John Newton: From 'On Female Dress'

There are two words in the Greek Testament, which are rendered *covetousness* in our version. The one literally signifies, *The love of money*; the other, *A desire of more.*

John Newton: From 'On Covetousness'

Interior of St. Mary Woolnoth – prior to alteration

I have somewhere met with a passage of ancient history; the substance of which, though my recollection of it is but imperfect, I will relate, because I think it very applicable to this part of my subject. It is an account of two large bodies of forces which fell in with each other in a dark night. A battle immediately ensued. The attack and resistance were supported with equal spirit. The contest was fierce and bloody. Great was the slaughter on both sides, and on both sides they were on the point of claiming the victory; when the day broke, and, as the light advanced, they soon perceived, to their astonishment and grief, that, owing to the darkness of the night, they had been fighting, not with enemies as they had supposed, but with friends and allies. They had been doing their enemies' work, and weakening the cause they wished to support. The expectation of each party to conquer the other, was founded upon the losses the opponents had sustained; and this was what proportionately aggravated their lamentation and distress, when they had sufficient light to show them the mischief they had done. Ah! my friends, if shame be compatible when the heavenly state, as perhaps, in a sense, it may, (for believers, when most happy here, are most sensibly ashamed of themselves), shall we not, even then, be ashamed to think how often, in this dark world, we mistook our friends for foes; and that, while we thought we were fighting for the cause of God and truth, we were wounding and worrying the people whom he loved, and, perhaps indulging our own narrow, selfish, party prejudices, under the semblance of zeal for his glory?

From a sermon preached in the Parish Church of St. Mary Woolnoth of the Nativity on Wednesday 21st November 1787, the day of the Annual Meeting of the Society for Promoting Christian Knowledge among the Poor given by the Rector, Rev. John Newton

Our beloved king is now on his way, amidst the acclamations of an affectionate people, to St. Paul's Cathedral; there he will, this day, make his public acknowledgement to God, who heard his prayer in the time of his trouble. It will be a joyful sight to thousands; and, perhaps, there is not a person in this assembly who has not felt a desire to be one of the spectators. But I am glad to meet you here. Many of you, I doubt not, earnestly and repeatedly prayed for the recovery of our gracious sovereign; and you judge with me, that the most proper expression of our gratitude and joy, is to unite in rendering prayer to God upon the very spot where we have often presented our united prayers. And I infer from the largeness of the congregation, that few who statedly worship with us are now absent; those excepted, who, residing in or near the line of procession, could not attend with propriety, nor perhaps with safety.

From 'The Great Advent', a sermon preached in the Parish Church of St. Mary Woolnoth of the Nativity on 23rd April 1789 on the occasion of the King's (George III's) happy recovery.

In the autumn of 1788, without telling her husband, Mar Newton sought the advise of a surgeon asking him to operat to cure her cancer. He found it inoperable and advised her t live quietly and smother the pain as best she could wit laudanum. When John returned home, he was shattered b the news and his faith was at first sorely tested. He did all h could for his dying wife, encouraging her with dainty morse of food.

He continued however as his wife lingered on with his paris duties. On the 23rd April 1789, he preached on the day o thanksgiving for the King's happy recovery. Not everyon would have been ready to pay a handsome tribute to th King, whose record of bribery and autocratic action wa legion. On the occasion of George III's recovery from thi particular bout of lunacy, he chose I Thessalonians 4, 16 an 17 for his text.

John Newton, despite his wide concerns about the needs of the poor and the enslaved, was happiest in the role of meeting individual need. *I seem to see in this world two heaps, of human happiness and human misery: now if I can take but the smallest bit from one heap, and add it to the other, I carry a point. If, as I go home, a child has dropped a halfpenny, and if, by giving it another, I can wipe away its tears, I feel I have done something. I should be glad, indeed, to do greater things: but I will not neglect this. When I hear a knock at my study-door, I hear a message from God. It may be a lesson of instruction, perhaps a lesson of patience: but since it is his message, it must be interesting.*

On 15th December 1790, Mary died. In the funeral address he preached on that text in Habakkuk which had inspired the final verse of his friend Cowper's *Sometimes a light surprises.* Just now, a great light had gone out of John's life. He found solace in the business of preaching and writing. Wilberforce had not moved Parliament but he was not one to give up. He had Newton's continued support. Cowper's poem *The Negro's Complaint*, printed on the finest paper and inscribed *A Subject for Conversation at the Tea Table*, was circulated everywhere, and even set to music. Josiah Wedgwood backed up Cowper's plea with a cameo of a slave in an attitude of pious entreaty. The anti-slavery lobby kept up the pressure.

In June 1791, the church of St. Mary Woolnoth was closed for repairs for four months. Newton visited many old friends and travelled to Bedford, to Teston in Kent, staying with Sir Charles and Lady Middleton *(kindness within doors; beautiful walks in the park)*, Dover, Bath and Bristol, and stayed with Hannah More at her home in Cowslip Green.

St. Mary Woolnoth

Rev. and dear Taureau,
The nice half pig (not the half nice pig) which you sent arrived safely. Coming from you, it was sure of a welcome. Thank you for pig and letter, and for every token and expression of your love to old friends.

The opening of a letter from John Newton to William Bull, January 1790

I considered her as a loan, which He who lent her to me had a right to resume whenever he pleased.

. . . . .

The Bank of England is too poor to compensate for such a loss as mine.

. . . . .

The good word of God was her medicine and her food, while she was able to read it. She read Dr. Watts' psalms and hymns, and the Olney hymns, in the same manner. There are few of them in which one, two, or more verses are not marked; and in many, which I suppose she read more frequently, each verse is marked.

John Newton, concerning his wife's last illness

How wonderful must be the moment after death! What a transition did she then experience! She was instantly free from sin, and all its attendant sorrows, and, I trust, instantly admitted to join the heavenly choir.

John Newton, following his wife's death

But thanks be to God, we have had, and still have both officers and privates, in the navy and in the army, whose courage is animated by Christian principles. They are not only defenders of their king and country, but are the servants of the Lord of hosts.

From John Newton's preface to an account of Lieutenant Colonel John Blackader's life published in 1799

Newton continued to receive a mountain of mail. *I have about sixty unanswered letters, and while I am writing one I usually receive two; so that I am likely to die much in debt.*

Newton's niece Betsy Catlett took him to see Cowper and Mary Unwin but the occasion was too poignant with past memories to be a success: Betsy was a good companion taking him to visit friends, but she too had indifferent health (at one time she had to spend time in Bedlam) but she recovered enough to continue looking after her aged uncle.

We were in the trenches all night, and only lost two men. I praise the Lord for preserving and defending me. Let others take it for chance or random, I look to a higher hand.

From the diary of John Blackader for 11th September 1706

William Bull was a great comfort to him too at this time, and they would often spend Saturday mornings together. In 1793, John Newton was introduced to Dr. Thomas Ring and his wife. Dr. Ring practised as a medical man in Reading and was to play a large part in securing the building of the Royal Berkshire Hospital. Whilst especially attached to the Church of England, the good doctor took a lively interest in all Christian work and the needs of the poor in Reading: the old widower found the Rings welcoming hosts and congenial company and between visits they enjoyed discussion by correspondence. Newton had always been ready to assist younger clergy as they worked out their ministry. The Reverend James Coffin held the living of Linkinhorne in Cornwall. Both he and Mrs. Coffin wakened to the importance of *vital religion* and sought Newton's help. Again he carried on his ministry by correspondence warning them about false humility. In a delightful thank you for a turkey, he sees it as coming from God like all the other blessings he received.

The *Pilgrim* is a parable, but it has an interpretation in which you are nearly concerned. If you are living in sin, you are in the City of *Destruction*. O hear the warning voice! *Flee from the wrath to come*. Pray that the eyes of your mind may be opened, then you will see your danger, and gladly follow the shining light of the Word, till you enter by *Christ*, the straight gate, into the way of salvation.

From John Newton's preface to an edition of John Bunyan's 'A Pilgrim's Progress' published in 1797

His last labours included a preface to an edition of John Bunyan's *Pilgrim's Progress* and another to the life and letters of Colonel John Blackader. In 1799, Newton edited the memoirs of the late Reverend William Grimshaw, minister of Haworth, for the benefit of the Society instituted for the relief of poor, pious Clergymen of the Established Church residing in the country.

On 30th March 1800, he preached a sermon in the Church of the United Parishes of St. Mary Woolnoth and St. Mary Woolchurch-Haw, Lombard Street, London, before the Right Honourable the Lord Mayor of London, Aldermen and Sheriffs for the benefit of the Children of Langbourn Ward Charity School. The School had been founded in 1702 and the printed version of the sermon notes that since then its scholars were:

Put out apprentices	391
To sea-services	16
Taken out, or otherwise disposed of	743
Now taught in the school	50
	1200

The Trustees and Committee in returning their thanks to the Subscribers and Public at Large for their support of the School, advised they had reduced the number of boys and were about to admit girls to the school. When John Newton was asked if copies of his sermon could be printed to help raise funds for this cause he adds: *The preacher cannot publish this Sermon as an exact copy of what he delivered from the pulpit. Some interval passed before he was desired to print it. His recollection is much impaired by old age; and he had no notes to assist it: but the plan is the same . . . As it is, he commends the perusal to the candour of the reader, and the blessing of Almighty God.*

He looks very old, and has got exceedingly fat since I saw him last, but he is full of piety, holiness and heavenly-mindedness.

William Bull writing on John Newton in 1793

Newton used to say *Now let us find a nut to crack.*

William Bull describes Saturday conversations with his friend.

What discoveries have been made in geometry, natural history, and chemistry! What powers are displayed in architecture, sculpture, painting, poetry and music! But, with respect to the concerns of his immortal soul, and the great realities of the unseen world, man, by nature, is dead as a stone. The dead body of Lazarus was not more incapable of performing the functions of common life, than we, by nature, are of performing one spiritual act, or even of feeling one spiritual desire; till He, who, by his commanding word, raised Lazarus from the grave, is pleased, by the power of his Holy Spirit, to raise us from the death of sin unto a new life of righteousness. He who, we profess to believe, will one day come to be our judge, has assured us that, except a man be born again, he cannot even see the kingdom of God. He has no faculty suited to the perception of what belongs either to the kingdom of grace upon earth, or what is revealed of the kingdom of glory in heaven. The result of his closest reasonings and shrewdest conjectures upon these subjects leave him in utter ignorance and darkness. As no description can communicate an idea of sun-shine or the colours of a rainbow to a man born blind, so the natural man cannot discern the things of God, for, they can only be spiritually discerned.

From a sermon preached in the Church of the United Parishes of St. Mary Woolnoth and St. Mary Woolchurch-Haw, Lombard Street, before the Right Honourable the Lord Mayor, Aldermen and Sherriffs on 30th day of March, 1800, by John Newton, Rector, for the benefit of the children belonging to Langbourn-Ward Charity School.

Mr. Newton is very feeble, – had great difficulty to get out of the coach. I was obliged to lift him with all my strength. He was most affectionate and said he would not have come so far for many people, only for me. He wished me to come and dine with him to-morrow, to be there at nine, and stay till seven . . . Everybody else shakes his head, and laments that he preaches at all.

William Bull writing in 1805

My friend, my friend! and have we met again,
Far from the home of woe, the home of men;
And hast thou taken thy glad harp once more,
Twined with far lovelier wreaths than e'er before;
And is thy strain more joyous and more loud,
While circles round thee heaven's attentive cloud?

Oh! let my memory wake! I told thee so;
I told thee thus would end thy heaviest woe;
I told thee that thy God would bring thee here,
And God's own hand would wipe away thy tear
While I should claim a mansion by thy side,
I told thee so – for our Emmanuel died.

John Newton: 'The supposed Meeting of Cowper and Newton in Heaven'

The same year, William Cowper died. A last task undertaken by Newton was to transcribe a sketch of the life of William's brother, John, who had died many years earlier, and who was greatly respected by Newton.

He continued to the end to value contact with William Bull. Newton had considered himself old when his wife died – but he was to be a widower for seventeen years. There was a late addition to the household. Betsy Catlett married and the couple lived with the old gentleman. Mrs. Betsy Smith took for her husband an optician based at the Royal Exchange. As Newton's health failed, so did his memory. But he still insisted on preaching and, when he could, used his influence for the complete abolition of slavery. In 1806 he preached for the last time. The following year, on 21st December the final words of his hymn *One there is above all others*, sung to ALL SAINTS (German) and ST. LEONARD (Bach) could be fulfilled.

But when home our souls are brought
We shall love thee as we ought.

In a time of severe and continual pain, he smiled in my face, and said – *Brother, I am as happy as a king.* **And the day before he died, when I asked him what sort of a night he had had, he replied,** *a sad night, not a wink of sleep.* **I said** *Perhaps though your mind has been composed, and you have been enabled to pray. Yes,* **said he,** *I have endeavoured to spend the hours in the thoughts of God and prayer; I have been much comforted, and all the comfort I got came to me in this way.*

William Cowper: From 'A Sketch of the Life of Rev. John Cowper' transcribed in 1802 by John Newton from William Cowper's hand-written manuscript.

The story has a curious postscript. In the early eighteen nineties the South London Underground Railway Company was seeking powers from Parliament to extend its line to Islington in North London. It took some time for Parliament to be persuaded to back the Bill, although it was clear that it would eventually be successful. In consequence, the mortal remains buried round St. Mary Woolnoth of the Nativity and in its crypt were removed to other consecrated ground: those of John and Mary Newton were re-interred in the south-east corner of Olney churchyard in 1893.

When the time eventually came to sell the crypt to the Railway the church authorities maintained that the valuation should be as prime commercial land, the mortal remains having been removed and, especially as if need be, St. Mary Woolnoth could be closed and its parish combined with another. The Railway, not surprisingly, challenged this claim right through to the House of Lords: but in the end had to part with £340,000, sufficient to build thirty new churches away from the centre of the City. The unwilling fairy god mother of many a London church may therefore be described as to-day's Northern Line of London's Underground.

St. Mary Woolnoth was not however closed. The church remains at the heart of London's business community with a special role of providing a ministry of healing to those who suffer from stress and tension. The parson who cared for William Cowper through mental breakdown would surely approve.

Should you find yourself in the booking hall of Bank (= Bank of England) Station, overburdened and distressed, think of its association with John Newton who wrote of the name of Jesus:

> It makes the wounded spirit whole
> And calms the troubled breast . . .

Bank Station booking hall - former site of crypt

Montgomery's birthplace at Irvine

When at Irvine I remember being remarkably struck with the full moon rising over the hills, and especially with her red appearance; and on another occasion, when she was in the form of a crescent, at a considerable height, only a few days old. There was on one occasion a great flood in the river, which did considerable damage to property, and powerfully impressed me, while I saw its waters rolling onward to the ocean. Inaminate nature was not the only thing that attracted my attention. On King George's birthday the people threw open their windows, while the soldiers fired over the houses: this produced a martial spirit; I got my little drum, and resolved to be a soldier.

James Montgomery writing in 1841 recalls early childhood at Irvine

Eternity, what are thou, – say?
– Time past, time present, time to come, – ***to-day.***

James Montgomery: From 'Questions and Answers'

The loud Atlantic ocean,
On Scotland's rugged breast,
Rocks, with harmonious motion,
His weary waves to rest,
And gleaming round her emerald isles,
In all the pomp of sunset smiles.
On that romantic shore
My parents hail'd their first-born boy: . . .

James Montgomery: From 'Departed Days'

James Montgomery

Nicholas Lewis, Count of Zinzenorf and Pottendorf, earned the title of *Renewer of the Old Church of the Brethren*, as, in the first half of the eighteenth century, the Moravian Church was reborn. In 1749, it was recognised in Britain by Act of Parliament as an ancient Protestant Episcopal Church in Great Britain and the Colonies. The Count reminded the Brethren that Christ alone was the Chief Elder of the Church which was guided, not by any human authority, but by the spirit of Truth received through faith and prayer. Moravians acknowledged there could be no Christianity without fellowship, both with Christ and with one another. The Count encouraged the many Moravian communities who established Brethren's Houses, Sisters' Houses, Clubs and Classes, to continue membership of and enrich other established churches, especially the Lutheran Church: but some Moravian communities did set up their own separate congregations.

The Montgomery family, of Scottish origin, had for some centuries been landed gentlefolk in Ulster; but James' great-grandfather had squandered the family's wealth. John Montgomery, James' father, was born at Ballykenny in County Antrim in 1733, the son of a labourer. In 1746, John Cennick, a celebrated Methodist preacher, came to live in the village, and joined the Moravians who set up a community at a place they named Grace Hill. John Montgomery was attracted to this Moravian group, who chose him to be a preacher and sent him to Germany and to Yorkshire. When he returned, he married Mary Blackley, a sister of the Society and shortly before James Montgomery was born on 4th November 1771, his father was appointed as Minister of the Moravian Church in Scotland at Irvine on the Firth of Clyde. James had two younger brothers, Robert who eventually set up a grocery business in Woolwich, and Ignatius who became a Moravian Minister. His only sister died in infancy.

In 1775 the family returned to Ireland: then in 1783 James' parents sailed to serve as missionaries to the negro slaves in Barbadoes, the family being placed in the care of the Moravian Community at Fulneck, (named after a Moravian town), near Leeds. This settlement consisted of three farms, with one building used as a Brethren's House, one as a Sisters' House, and the third as a School. John and Mary Montgomery hoped their eldest son would become a Minister, but he was none too willing a pupil, being more interested in penning poetry than the subjects on the school timetable. His interest was heightened in this direction when he read a poem by Robert Blair titled *The Grave* and Richard

We never enter into controversy with any other denomination, nor do we endeavour to draw their members over to us.

. . . . .

We never attempt, by means of our missions, to obtain the least influence in civil or commercial affairs, but are contented with what we can earn by our industry in useful employment for our support, to the satisfaction of the government.

. . . . .

We carefully avoid intermeddling with anything that can increase the wrong and prejudicial ideas which the Heathen, Savages or Slaves have imbibed against the Christian religion.

. . . . .

We confess and preach to the Heathen, *Jesus Christ and him crucified* as the Saviour of the world, because *there is no other name under heaven given among men whereby we can be saved, but the name of Jesus Christ*; and we seek, as far as in us lies, to keep them ignorant of the many divisions in Christendom: But if they happen to have been informed thereof by others, we endeavour with great precaution to approve ourselves impartial, speak of the several divisions with much tenderness, and to extenuate, and not exaggerate the differences, that thus the knowledge of the mystery of Christ may be increased, and misapprehensions diminished.

Augustus Gottlieb Spangenberg: From 'A Candid Declaration of the Church known by the Name of the Unitas Fratrum, relating to their labour among the Heathen'

Blackmore's *Prince Arthur.*

The new-made widow too, I've sometimes spy'd,
Sad sight! slow moving o'er the prostrate dead:
Listless, she crawls along in doleful black,
Whilst bursts of sorrow gush from either eye,
Fast falling down her now untasted cheek.
Prone on the lowly grave of the dear man
She drops; whilst busy-meddling memory,
In barbarous succession, musters up
The past endearments of their softer hours,
Tenacious of its theme. Still, still she thinks
She sees him, and, indulging the fond thought,
Clings yet more closely to the senseless turf,
Nor heeds the passenger who looks that way.

Robert Blair: From 'The Grave'

I shall not easily forget the boys' sleeping hall, a large room which extended over the whole building appropriated to the school, and contained between one and two hundred beds. It was usual for us to meet there on the evening prior to Easter Sunday. A pianoforte was taken for the occasion to one end of the immense room; over it was suspended a lantern, which threw a dim light on a splendid painting of a dead Christ, removed from the brethren's house. When all had assembled, we stood for a few minutes in front of the picture. Then the full-toned piano, accompanied by a French bugle, broke the silence with one of those airs which for ages have been used in the Moravian Church. This ceased for a moment, and we heard the sweet melody whispering round that vast hall, the whole of which was in darkness, save the spot where we were gathered. Again we mused on the painting, and were almost startled by the breathless quiet of the place.

James Montgomery describing Easter Eve at Fulneck in the 'Metropolitan Magazine'

Ambitious *Lucifer*, depos'd of late
From Bliss Divine, and high Angelick State,
Sinks to the dark, unbottom'd Deep of Hell,
Where Sin, and Death, and endless Sorrow dwell:
Here plung'd in Flame, and tortur'd with Despair
He plots Revenge, and meditates new War.
His Thoughts on deep Designs th'Apostate spent,
When this Conjuncture favour'd his Intent.
A spacious, dusky Plain lay wast and void,
Where yet Creating Power was ne'er imploy'd
To fashion Elements, or strike out Light;
The silent, lonesome Walks of ancient Night.
In th'Archives kept in Heav'n's bright Towers, was found,
A sacred old Decree, wherein the Ground
Was set distinctly out, from Ages past,
For a new World, on this unbounded Wast.
Here did th'Artificer Divine of late,
The World so long before markt out, create.
And gave it to the Man he newly made,
Where all things him, as he did Heav'n, obey'd.
In *Eden's* Walks he made his blest Abode,
All full of Joy, of Glory, full of *God.*
Nature with vast Profusion on him pours
Unmeasur'd Bliss, from unexhausted Stores.

Richard Blackmore: From Book 1 of 'Prince Arthur'

Meanwhile, soon after their arrival, a Mr. Hamilton sought a missionary for his estate in Tobago. The island was then in French possession and consequently under a Roman Catholic Government. Eventually the Synod at Herrnhüt agreed the Montgomerys should go to Tobago. Once they were there, the French soldiers mutinied and set fire to the town.

James' guardians were not sure what to do with their teenage charge: clearly he did not have the makings of a Moravian Minister. They placed him with a Mr. Lockwood, a Mirfield baker, to learn the trade. James was certain that bread-making was not his calling either: he used every spare moment to scribble poetry or to act music-mad blowing his *brains out with a hautboy* in a neighbouring hayloft. On Sunday, 19th June 1789, he was missing from the preaching meeting of the local Moravian Brethren: the seventeen year old had set out, while his employer was at worship, in an old suit, carrying a single change of other clothing, and three shillings and sixpence. He left behind a new suit recently bought for him by Mr. Lockwood, feeling he had done insufficient work to justify his keeping it.

The next morning found us assembled at five o'clock in the chapel, joined by an immense crowd. The service opened with a voluntary on the organ, followed by the Litany, the responses of which were sung by the choir and congregation. On arriving at the part which refers to the Church triumphant, all adjourned to the burial-ground, and there finished the service in the open air.
Those only who have witnessed it can form any notion of its solemnity. The congregation formed a circle, in the centre of which was the officiating clergyman. The sun had just risen, and was lighting up that splendid scenery, and the mists of the night were rapidly rolling away. In the distance, covering the opposite hill, were magnificent woods, swept by a clear crystal stream; over us the birds of the morning carolled their early matins, and then soared into high heaven.

James Montgomery describing Easter Morning at Fulneck in the 'Metropolitan Magazine'

Time, whither dost thou flee?
— I travel to Eternity.

James Montgomery: From 'Questions and Answers'

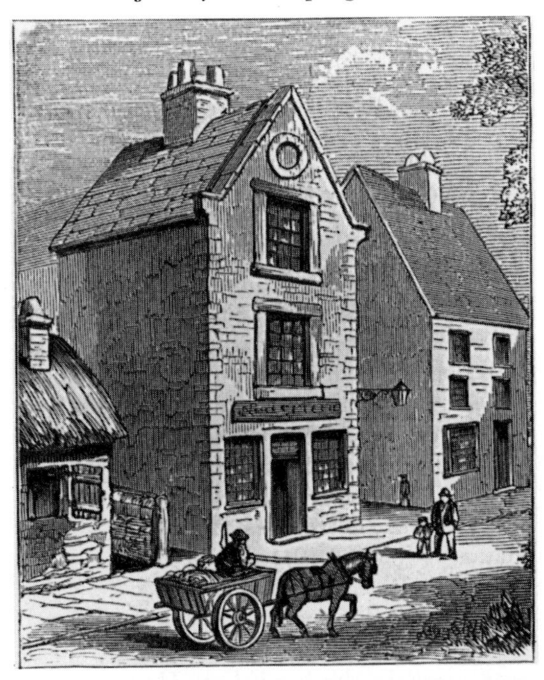

The house in which Montgomery lived at Wath upon Dearne

Two days later, James lodged at a small public house in a snug little village near Wentworth where the likeable young man was apparently allowed to stay for several days free of charge. Knowing he must earn something, he heard of a vacancy at Mr. Joshua Hunt's emporium at Wath upon

Dearne. To secure the job, he needed references from his old employer who at the request of the Fulneck community visited the lad; and, finding him set upon making his own way in the world, gave him the necessary good character. James' winning ways again stood him in good stead when, meeting with Earl Fitzwilliam, he set out to sell him some of his poems and ended up with a guinea for his work.

Ye Dead, where can your dwelling be?
– The house for all the living: – come and see.
James Montgomery: From 'Questions and Answers'

Meanwhile his parents had suffered great hardships. Not long after the French mutiny, Mary Montgomery was sick of a violent fever. After three anxious days and nights with a physician in frequent attendance, a storm rose to a hurricane. An unoccupied tumbledown house next door was thrown upon the missionaries' home and they were lucky not to be buried alive. As it was, John Montgomery wrote that *he carried* his *poor wife into an adjoining chamber*; but though it was *very firmly built, the rain beat in at all corners, so there was but one spot where* his *wife could sit dry. In this situation,* they *waited till the storm abated, and were graciously preserved from further harm, excepting that* his *wife's illness increased, and* he *got so violent a cold that* he *did not recover within a fortnight* after. Mary's fever returned after a few weeks, this time proving fatal and her husband succumbed to fever during the following year. The mission work in Tobago was to lay dormant from 1791 to 1825 when Mr. Hamilton bequeathed a considerable legacy for its continuance and the ladies of Sheffield presented two hundred sovereigns to the then well-known citizen James Montgomery for the work.

Sweet seas and smiling shores!
Where no tornado-demon roars;
Resembling that celestial clime
Where with the Spirits of the Blest
From all their toils my parents rest.
James Montgomery: From 'Departed Days'

But back to the story of young James who was encouraged by the local bookseller in Swinton, Mr. Brameld, to continue his efforts as a poet; the business of selling flour and boots and pots and pans was very much a means to other ends. In due time, Mr. Brameld agreed to send a batch of poems to Mr. W. Harrison, a publisher and bookseller in London's Paternoster Row. James, anxious to ensure his verses had every chance, gave in his notice and followed his poems to London. He was unsuccessful in persuading Mr. Harrison to publish them but the young traveller was again fortunate in being offered a job in the bookshop. He used all his off-duty hours pedalling his poetry until, after a year, he was forced to admit to himself he would not find a publisher. He recounted later to friends that when retreating from one bookseller, who like so many had crushed his aspirations, he dashed his head against a patent lamp, broke the glass and spilled the oil amid the titterings of the shopkeeper behind the counter.

So the prodigal poet, now without funds, returned to Wath upon Dearne, and was welcomed back by Joshua Hunt and again given employment. Indeed, he was treated more like a son than an employee: and Mrs. Hunt was greatly moved when he wrote verses for her when she suffered a very painful affliction. In Wath, he met Hannah Turner with whom he walked out on several occasions. In March 1792, twenty year old James was collecting debts for his employer at Great Houghton, near Darford, when he saw an advertisement for a clerk in the *Sheffield Register*. He applied to the paper and got the job.

I came to this town in the Spring of 1792, a stranger, and friendless, without any intention or prospect of making a long residence in it . . .
James Montgomery, in a farewell speech in 1825

At fond sixteen my roving heart
Was pierced by Love's delightful dart:
Keen transport throbb'd through every vein,
– I never felt so sweet a pain!

Where circling woods embower'd the glade,
I met the dear romantic maid:
I stole her hand, – it shrunk, – but no.
I would not let my captive go.

With all the fervency of youth,
While passion told the tale of truth,
I mark'd my Hannah's downcast eye –
'Twas kind, but beautifully shy:

Not with a warmer, purer ray,
The sun, enamor'd woos young May;
Nor May, with softer maiden grace,
Turns from the sun her blushing face.

But, swifter than the frighted dove,
Fled the gay morning of my love;
Ah! that so bright a morn, so soon,
Should vanish in so dark a noon.

The angel of Affliction rose,
And in his grasp a thousand woes;
He pour'd his vial on my head,
And all the heaven of rapture fled.

Yet, in the glory of my pride,
I stood, – and all his wrath defied;
I stood, – though whirlwinds shook my brain,
And lightnings cleft my soul in twain.

I shunn'd my nymph; – and knew not why
I durst not meet her gentle eye;
I shunn'd her – for I could not bear
To marry her to my despair.

Yet, sick at heart with hope delay'd,
Oft the dear image of that maid
Glanced, like a rainbow, o'er my mind,
And promised happiness behind.

The storm blew o'er, and in my breast
The halcyon Peace rebuilt her nest:
The storm blew o'er, and clear and mild
The sea of Youth and Pleasure smiled.

'Twas on the merry morn of May,
To Hannah's cot I took my way:
My eager hopes were on the wing,
Like swallows sporting in the Spring.

Then as I climb'd the mountains o'er,
I lived my wooing days once more;
And fancy sketch'd my married lot,
My wife, my children, and my cot.

I saw the village steeple rise, –
My soul sprang, sparkling in my eyes:
The rural bells rang sweet and clear, –
My fond heart listen'd in mine ear.

I reach'd the hamlet: – all was gay;
I love a rustic holiday;
I met a wedding, – stepp'd aside;
It pass'd, – my Hannah was the bride.

– There is a grief that cannot feel;
It leaves a wound that will not heal;
– My heart grew cold, – it felt not then:
When shall it cease to feel again?

James Montgomery: 'Hannah'

The newspaper was a small four-page weekly with a circulation of around two thousand copies. Its proprietor-editor was Joseph Gales, whose business was also that of general printer, stationer, book-seller and auctioneer. He also held strong political anti-Government views, and used his paper as a propaganda vehicle for causes dear to him. He particularly appreciated the writings of Thomas Paine whose *Rights of Man* was published in 1791.

[Burke] is not affected by the reality of distress touching his heart, but by the showy resemblance of it striking his imagination. He pities the plumage, but forgets the dying bird.

Thomas Paine: From 'Rights of Man'

At Sheffield, young James found a friend and counsellor in Ebenezer Rhodes, later to become well known as a travel writer. James began work in the counting-house dealing with auctioneering transactions but quickly was writing pieces for the paper, with whose radical views he found himself in sympathy. The need for parliamentary reform was paramount but the Republicanism of France was not to be followed. An early article signed J.M. is headlined *Modes of Government*. Under the pseudonym Gabriel Silvertongue, James put out a satirical item *The History of a Church and a Warming Pan* written for the benefit of the Associates and Reformers of the Age and dedicated, without permission, to their tri-fold majesties, the People, the Law and the King. In the story Sunday collections made in the warming pan for urgent repairs were spent at the local ale house during the week. There was urgent need of ecclesiastical as well as political reform.

Sir! – May your Majesty be the greatest Monarch, and your subjects the greatest people, in the universe; and, long ere you descend to your forefathers, may your Majesty see restored to all your children their birthright – equal Privileges and equal Laws.

James Montgomery: From the 'Dedication to the King's most excellent Majesty' in 'The History of a Church and a Warming-Pan'

The Bull Inn at Wicker provided a regular meeting place for Proprietor Gales, Charles Sylvester, later a renowned civil engineer, Ebenezer Rhodes and young James. Here they put the world to rights week by week! The Mayor and Corporation made much of their loyalty in subscribing fifty guineas for comforts for the troops fighting the French. The Bull Inn conspirators decided to play a little joke on the pompous mayor, George Pearson. They ensured he received a letter purporting to offer him a knighthood in recognition of this generous gesture. The Mayor reported the matter to the Corporation only to discover it was all a hoax. *The Sheffield Register* carried a 136-line poem, *The Mayor of Donchester* by Jonathan Starlight alias James Montgomery.

The House of Correction may with propriety be compared to Cobbler's shop: the oftener a pair of old shoes are repaired the *worse* they are for mending; and the more some offenders are *corrected*, the more *incorrigible* they become!

From 'Sheffield Register' of 16th August 1793

'Fools, fools apiece!' exclaimed a fat
Old wag, who in a corner sat.
'Fool?' cried the Mayor, enraged, confounded:
'Fools?' roared the Aldermen astounded.
'Yes, fools apiece! fools altogether!'
Replied the wicked sneering wag.

'Fools! Fools!' he cried, 'Fools of a feather!
I'll let the cat out of the bag;
This letter, sirs, was never wrote by Grenville,
But forged, I ween, on quite another anvil;
Some Jacobin has coined this fabrication,
Just to befool our learned Corporation.'

From 'The Mayor of Donchester' by 'Jonathan Starlight' in the 'Sheffield Register' of 21st February 1794

We are happy to have it in our power at any time to record acts of equity, and we know of none that better deserves to be handed down to posterity, than the following recent one, which is an example worthy the imitation of the administrators of that great blessing – *Justice!* – A poor weaver, in the neighbourhood of Manchester, on being inlisted for a soldier, was carried before a Rev. Magistrate, to be sworn in; the man, on having the oath rendered him, hesitated, and bursting into tears, was asked what objections he had to entering into his Majesty's service after having enlisted, without being *cajoled*? The poor fellow in reply said, that he had brought that morning a piece of calico to town, but, that instead of receiving his wages, he was threatened with a warrant, for having *spoiled* his work, and turned out of the warehouse *pennyless!* – that, knowing the deplorable state of his family (his wife lying-in) and thinking there was no prospect of gaining redress from *so respectable a house*, tho' he knew his *piece* to be well worked (for added he *these are no times for spoiling one's pieces*) he had determined to inlist, and, with the bounty-money that he might receive, supply their *present* wants, and to leave their *future* support to Providence!! – His unvarnished tale being ended, had a sudden effect on all present, and, bringing to recollection many similar cases – highly to the honour of the Justice, he was determined to inquire into the truth of the story, and, if just, to see the poor man righted: a verbal message was sent to the house, which being disregarded, *summons* of course followed, and one of the partners of the house attending, the piece was produced, re-examined, and being found *well* wrought, the wages were immediately paid down, and the weaver suffered to return home (without paying *smart*) to his expecting wife and *seven helpless children*! to whom we hope he will long continue a support! – We cannot close this article, without execrating the conduct of any man, or set of men, who, by their own, or servants' acts of injustice, deprive the country of its *most useful members*, by *forcing* them from their friends and their families, to seek a support in distant climes, amongst the din of war, and where little is to be *gained* save *grinning honour.*

From 'Sheffield Register' of 11th April 1794

. . . he was a real politician long before he was a true poet.

J. W. King: From 'James Montgomery'

It is lovely then to watch the moonbeams penetrate the deep recesses of the dale – light up the foliage on the brink of the river, and, gliding through the over-hanging branches, spread over all the bosom of the stream that soft, yet mellow lustre, which moonlight only can bestow. The general stillness which at this time pervades the dale, gives a peculiar charm to the gentle rush of water, as it passes over the weir near the mills below.

Ebenezer Rhodes: From 'The Derbyshire Tourist's Guide and Travelling Companion' in which the author describes scenery at Matlock Bath

You were no doubt astonished when first you saw my name annexed to the *Iris*; and perhaps still more when you observed the humiliating distance between the cringing, trembling, gouty pace of our party-coloured messenger of the gods, and the noble, firm, and manly gait of the late lamented *Register* . . . I cannot expect that the *Iris* will meet with the liberal patronage which supported the late *Register*.

James Montgomery writing to a friend in 1794

Ours are the plains of fair delightful peace,
Unwarped by party rage, to love like brothers.

The motto of the newly-launched 'Iris'

Before the year was out, Joseph Gales had to flee the country to escape prosecution and eventually reached Philadelphia where he continued his life as a journalist. His two sisters continued to live at Hartshead and James Montgomery was to have his home there too for many years. In 1839, when Joseph Gales was in his seventy-ninth year and the fifty-fifth year of his marriage, he wrote on learning that his sister Anne had died, of Montgomery having become a substitute brother. At first, James often walked to Wath to see his girl friend: but they fell out and on a later visit he was to find her at an altar marrying another. He remained a bachelor all his life. With the help of a Unitarian minister, the Reverend Benjamin Naylor of Upper Chapel in Norfolk Street, funds were found to continue the paper: but under its new name *The Iris*. It was given a new motto which renounced, to the sadness of Montgomery, its former radicalism. The authorities were looking for an excuse to close the *Iris*, but its columns failed to give them an opportunity.

However, in the general printing business, the firm ran off at the request of a customer copies of a ballad earlier published in the *Register*, a patriotic song by a clergyman of Belfast.

When the constable arrested James for seditious libel respecting the war then raging between His Majesty and the French Government, he was completely puzzled. When the ballad was produced, the fact that it had been written before hostilities even began between England and France, and actually related to another historical event, was of no avail.

The Jury were to consider the intention only; for neither the printing nor the publication, if they were ever so clearly proved, could constitute the guilt or innocence of the defendant, but the design and intention alone.

From the Judge's summing up of the trial of James Montgomery at Doncaster Sessions

The magistrate committed him for trial at Doncaster Sessions and the court was crowded on 22nd January 1795 for the nine-hour hearing. The jury, after being locked up nearly an whole hour, returned with a verdict *Guilty of Printing and Publishing*, which was not an answer to the charge in the indictment. They were sent out again and after another fifty minutes, found Montgomery guilty. Despite a plea from his former employer Joshua Hunt he was fined and sentenced to three months imprisonment which he served in York Castle.

The magistrate passing sentence told him that as a young man, and for an offence like his, he might think himself well-off he was not ordered to stand in the pillory for an hour. On 11th February the Society of Friends of Literature voted an address to him. He wrote *Verses to a Robin Redbreast* in honour of a regular visitor to his cell window.

Welcome, pretty little stranger!
Welcome to my lone retreat!
Here, secure from every danger,
Hop about, and chirp, and eat:
Robin! how I envy thee,
Happy child of Liberty!

James Montgomery: From 'Verses to a Robin Red-breast, who visits the Window of my Prison every Day'

The King v James Montgomery

That James Montgomery, late of Sheffield aforesaid, in the Riding and County aforesaid, Printer, being a wicked, malicious, seditious, and evil disposed person, and well knowing the premises, but wickedly, maliciously, and seditiously contriving, devising, and intending to stir up and excite discontent and sedition among his Majesty's subjects, and to alienate and withdraw the affection, fidelity, and allegiance of his said Majesty's subjects from his said Majesty, and unlawfully and wickedly to seduce and encourage his said Majesty's subjects to resist and oppose his said Majesty's Government and the said War, on the 16th day of August, in the thirty-fourth year of the reign of our said Lord the now King, with force and arms at Sheffield aforesaid, in the Riding and County aforesaid, unlawfully, wickedly, maliciously, and seditiously did print and publish, and cause and procure to be printed and published, a certain false, scandalous, malicious and seditious Libel, of and concerning the said War, and his said Majesty's conduct thereof, entitled *A Patriotic Song by a Clergyman of Belfast.*

Part of the Indictment preferred at Sheffield Michaelmas Sessions, 1794

Nothing can more clearly demonstrate the ignorance and virulence of the High Party on the other side of the Tweed than that the bail of a substantial citizen, a Burgess and Guild-Brother of Leith, was rejected last week by a wise magistracy, because, forsooth, he was guilty of reading — the *Iris*.

From the 'The Iris' of 17th July 1794

Not long after his release, the Reverend Benjamin Naylor, who was about to get married, sought to disengage himself from the hazardous business of newspaper ownership and the business was sold to James Montgomery for £1,760 to be paid over a period. Before long, however, he was in trouble again with the authorities, this time over an account printed of a local riot, which a rival newspaper, the *Sheffield Courant* had reported very differently. These were days when the authorities feared a general uprising and were concerned to suppress and hide any unrest. In court, charged with false reporting constituting a seditious libel, Montgomery found himself back in York Castle, this time to serve a six month sentence. A village preacher, John Pye Smith, graciously kept the paper going for him. Throughout his life,

The Divine realities of *regeneration* and *sanctification* by the Holy Spirit, meet the conscious weakness and all the wants and miseries of sinful man. Effectual provision is made, in the blessed gospel, for the implantation and growth of *real holiness* in the heart.

John Pye Smith: From 'A Sermon preached at the Monthly Meeting of the Society for the Education of Young Men for the Work of the Ministry among Protestant Dissenters'

We know the power of Monarchs to create and bestow titles. One of the most illustrious Orders of Knights in Great Britain derive the pedigree of their insignia from a lady's garter: a Roman Emperor, after triumphing over a prodigious army of cockle shells, which he utterly discomfited in a mighty battle on the shores of Gaul, intended to have conferred the Consulship upon his favourite horse: and one of our own facetious Princes rewarded a loin of beef with the honour of knighthood for its eminent *table services*: a mark of distinction which its successors have inherited ever since, the hinder quarter of an ox being to this day worshipped by all pious beef-eaters under the title of Sir Loin!

From 'Hints concerning Title-pages, and a Whisper about Whispering' written by James Montgomery under the pseudonym Gabriel Silvertongue in 'The Iris' of 5th June 1795

The man who first invented dinners
Was certainly the chief of sinners;
For those who once the habit gain,
May long to leave them off in vain:
Nor even in gaol can folk forget,
To eat, to drink, and run in debt!
Thousands, by dinners, are undone,
But woe to those who can get none!
Though many a one has died with dining,
Yet many more have perished pining:
While too much dinner is a curse,
No dinner is as bad, or worse;
But who would give a pin to choose
To die of famine or roast goose?
In this sweet place, where freedom reigns,
Secured by bolts and snug in chains;
Where innocence and guilt together
Roost like two turtles of a feather;
Where debtors safe at anchor lie,
From saucy duns and baillifs sly;
Where highwaymen and robbers stout,
Would, rather than break in, break out;
Where all's so guarded and recluse,
That none his liberty can lose! –
Here each may, as his means afford,
Dine like a pauper or a lord;
And he who can't the cost defray,
Is welcome, sir, to fast and pray!

James Montgomery writing under the pseudonym Paul Positive: From 'The Pleasures of Imprisonment': 'The Second Epistle to a Friend'

Many of the pieces were composed in bitter moments, amid the horrors of a gaol, under the pressure of sickness. They were the transcripts of melancholy feelings – the warm effusions of a bleeding heart. The writer amused his imagination with attiring his sorrows in verse, that, under the romantic appearance of fiction, he might sometimes forget that his misfortunes were real.

James Montgomery from his Preface to 'Prison Amusements' published in 1797 after his release from York Castle

Montgomery gathered round him a loyal band of friends: and these associations often flourished for many years. Even when Montgomery was released, his health had suffered and Pye Smith continued his duties so James could take time at Scarborough for recuperation. A volume of writings in prison was published under the title of *Prison Amusements*. A Quaker, Henry Wormwell, in prison at York wrote in his diary: *Went from this place James Montgomery, a very kind and social young man: he was to me a pleasing companion, and he has left a good report behind him. Although he is qualified with good natural parts, and has had a liberal education, yet he was instructive and kind to me. I think I never had an acquaintance with any one before, that was not of my persuasion, with whom I had so much unity, I was provided and thought it a loss to part with him.*

What bird, in beauty, flight, or song,
Can with the Bard compare,
Who sang as sweet, and soar'd as strong,
As ever child of air?

His plume, his note, his form, could *Burns*,
For whim or pleasure change;
He was not one, but all by turns,
With transmigration strange.

The Blackbird, oracle of spring,
When flow'd his moral lay;
The Swallow wheeling on the wing,
Capriciously at play:

The Humming-Bird, from bloom to bloom,
Inhaling heavenly balm;
The Raven, in the tempest's gloom;
The Halcyon, in the calm:

In *auld Kirk Alloway*, the Owl,
At witching time of night;
By *bonnie Doon*, the earliest Fowl
That caroll'd to the light.

He was the Wren amidst the grove,
When in his homely vein;
At Bannockburn the Bird of Jove,
With thunder in his train:

The Woodlark, in his mournful hours;
The Goldfinch, in his mirth;
The Thrush, a spendthrift of his powers,
Enrapturing heaven and earth;

The Swan, in majesty and grace,
Contemplative and still;
But roused, – no Falcon, in the chase,
Could like his satire kill.

The Linnet in simplicity,
In tenderness the Dove;
But more than all beside was he
The Nightingale in love.

Oh! had he never stoop'd to shame,
Nor lent a charm to vice,
How had Devotion loved to name
That Bird of Paradise!

Peace to the dead! – In Scotia's choir
Of Minstrels great and small,
He sprang from his spontaneous fire,
The Phoenix of them all.

James Montgomery: 'Robert Burns'

Robert Burns died that year. Like Montgomery, he was born at Irvine and, many years later, the surviving poet wrote memorial verses. Walking alone was always a popular pastime and his close observation of nature are evident in *The Wild Rose: on plucking One late in the Month of October* and *Verses, on finding the Feathers of a Linnet scattered on the Ground in a solitary Walk*. The next year the *Sheffield Courant* was abandoned: and Montgomery wrote the hymn *When, like a stranger on our sphere* for the opening of the Sheffield General Infirmary. He gave this cause great support over the years. In the hymn, he recalls the healing miracles of Jesus and sees the relief of misfortune, sickness and poverty as work to be done by his followers to-day.

. . . many of whom in their daily employments are peculiarly exposed to the hazard of contracting dangerous diseases, and of suffering from calamitous accidents.

From the sermon by Reverend James Wilkinson on the occasion of the opening of Sheffield Infirmary

Montgomery's poems in 1799 include *The Vigil of Saint*

Mark which is based on the curious legend of the Eve of the Feast.

The ghost of all whom Death shall down
Within the coming year,
In pale procession walk the gloom
Amid the silence drear.

In the poem, Edmund proposes to Ella who says she will only marry him if he witnesses the pale procession in which Edmund sees his girl in a shroud. James often revealed his sense of humour when writing for the *Iris*. In 1800 he reported: *The clocks of Copenhagen are so extremely well bred, that they never strike the hour till the Palace clock has given the signal.* That same year he wrote an address, spoken at Sheffield Theatre on the occasion of the performance of *The Jew* for the benefit of the poor by the gentlemen of the Thesbian Society, when England was suffering from a universal shortage of food. In introducing his poem *The Battle of Alexandria* written in 1801, Montgomery writes: *At Thebes, in ancient Egypt, was erected a statue of Memnon, with a harp in his hand, which is said to have hailed with delightful music the rising sun, and in melancholy tones to have mourned his departure.*

By an accurate calculation of European taxes, it appears that an Englishman pays as much as three Hollanders, five Spaniards, six Austrians, nine Portuguese, ten Frenchmen, twelve Turks, and fourteen Russians.

Editorial comment in 'The Iris' of 14th July 1797

But while we are prudently preparing to meet the worst of all possible calamities, a civil war in the bosom of Britain, let us not, for a moment, forget, that the return of tranquility alone can establish our independence as a nation, and our happiness as a people. The most splendid victories are only calamities in brilliant disguise: the proudest triumphs are funereal pageants, that should humble the living. For they insult the dead. The laurels of Conquerors are watered with tears, and flourish in a soil of blood; in vales of felicity, on mountains of repose, spring the vineyards of plenty and the olive groves of peace.

From an editorial in 'The Iris' in April 1798

In spring to build thy curious nest,
And woo thy merry bride,
Carol and fly, and sport and rest,
Was all thy humble pride.

James Montgomery: From 'Verses on finding the Feathers of a Linnet scattered on the Ground, in a solitary Walk'

Harp of *Memnon*! sweetly strung
To the music of the spheres;
While the *Hero's* dirge is sung,
Breathe enchantment to our ears.

. . . . .

Let thy numbers, soft and slow,
O'er the plain with carnage spread,
Soothe the dying while they flow
To the memory of the dead.

James Montgomery: From 'The Battle of Alexandria'

Here the whole family of woe
Shall friends, and home, and comfort know;
The blasted form and shipwreck'd mind,
Shall here a tranquil haven find.

James Montgomery: From 'Hymn for the Opening of Sheffield General Infirmary, October 1797', beginning 'When, like a stranger on our sphere'

Around thy bell, o'er mildew'd leaves,
His ample web a spider weaves;
A wily ruffian, gaunt and grim,
His labyrinthine toils he spreads
Pensile and light; – their glossy threads
Bestrew'd with many a wing and limb;
Even in thy chalice he prepares
His deadly poison and delusive snares.

James Montgomery: From 'The Wild Rose; on plucking One late in the Month of October'

This shadow on the Dial's face,
That steals from day to day,
With slow, unseen, unceasing pace,
Moments, and months, and years away;
This shadow, which, in every clime,
Since light and motion first began,
Hath held its course sublime; –
What is it? – Mortal Man!
It is the scythe of *Time*:
– A shadow only to the eye;
Yet, in its calm career,
It levels all beneath the sky;
And still, through each succeeding year,
Right onward, with resistless power,
Its stroke shall darken every hour,
Till Nature's race be run,
And *Time's* last shadow shall eclipse the sun.

James Montgomery: From 'The Dial'

Lyre! O Lyre! my chosen treasure,
Solace of my bleeding heart;
Lyre! O Lyre! my only pleasure,
We must now for ever part;
For in vain thy poet sings,
Woos in vain thine heavenly strings;
The Muse's wretched sons are born
To cold neglect, and penury, and scorn.

James Montgomery: From 'The Lyre'

The *Sun* is but a spark of fire,
A transient meteor in the sky;
The *Soul*, immortal as its Sire,
Shall never die.

James Montgomery: Closing verse of 'The Grave'

Since our last, four Milk Sellers have been convicted before the Magistrates, and paid the penalty of making short measures. The public pay dear enough for this necessary without being abridged in the quantity: – we wish the Magistrates could persuade the Vendors to improve the quality of their article, by mixing a little more Milk with the Water.

Editorial comment in 'The Iris' of 19th June 1800

Minutes of Quarterly Board of Governors of Sheffield Infirmary, 23rd March 1801

2 . . . That therefore it will be absolutely necessary, either to increase the receipts or diminish the number of patients.

3 That, to avoid an alternative so painful to humanity as the latter, a Committee of Subscriptions . . .

Among the Benevolent Institutions in this Town, for the benefit and improvement of the situation of the Poor, the School for Industry ought to be distinguished, as a plan excellently adapted for the introduction of female children in the few branches of knowledge, necessary for their comfort, and for teaching them habits of cleanliness, diligence and honesty. The Institution is still young and weak, and demands a little of the sunshine of public favour.

Editorial comment in 'The Iris' of 17th September 1801

Though our Parent perish'd here,
Like the Phoenix on her nest,
Lo! new-fledg'd her wings appear,
Hovering in the golden West.

Thither shall her sons repair,
And beyond the roaring main
Find their native country there,
Find their *Switzerland* again.

Mountains, can you chain the will?
Ocean, canst thou quench the heart?
No; I feel my country still,
Liberty! where'er thou art.

Thus it was in hoary time,
When our fathers sallied forth,
Full of confidence sublime,
From the famine-wasted North.

Freedom, in a land of rocks
Wild as Scandinavia, give,
Power Eternal! – where our flocks
And our little ones may live.

Thus they pray'd; – a secret hand
Led them, by a path unknown,
To that dear delightful land
Which I yet must call my own.

To the Vale of *Schwitz* they came:
Soon their meliorating toil
Gave the forests to the flame,
And their ashes to the soil.

Thence their ardent labours spread,
Till above the mountain-snows
Towering beauty show'd her head,
And a new creation rose!

– So, in regions wild and wide,
We will pierce the savage woods,
Clothe the rocks in purple pride,
Plough the valleys, tame the floods; –

Till a beauteous inland isle,
By a forest-sea embraced,
Shall make Desolution smile
In the depth of his own waste.

There, unenvied and unknown,
We shall dwell secure and free,
In a country all our own,
In a land of Liberty.

James Montgomery: From 'The Wanderer of Switzerland'. Not long after publication, an American edition of 1000 copies was published by George Bourne in New York.

The wind here rose early in the morning and blew with increasing violence till late in the afternoon. Scarcely a house in the town escaped dilapidation . . . a girl about twelve years old was crushed to death at Sheffield Moor, by the ruins of a wall, which was thrown upon her, as she endeavoured to support herself against it in the wind . . . a large spinning factory, at Pendleton, was blown down . . .

News of the great storm acrosss the country filled many columns of 'The Iris' of 28th January 1802

His poems for 1802 include one on the mystery of Time called *The Dial.* In 1803, he wrote a column in the *Iris: The heart of Switzerland is broken! and liberty has been driven from the only sanctuary which she found on the continent. But the unconquered and unconquerable offspring of Tell, disdaining to die slaves in the land where they were born free, are emigrating to America* . . . Three years later he wrote a lengthy narrative poem about a Swiss refugee, telling his story to a shepherd giving him and his family sanctuary as they fled the French invaders and hoped for a new life in America. There was a tradition that long ago, the Swiss had left a Scandinavian homeland in time of famine under an edict that a tenth of the population must emigrate lest all starve to death. In the poem, *The Wanderer of Switzerland,* he links this ancient tradition with the new emigration. Other oetry from the early years of the nineteenth century ncludes *The Lyre* in which words were given to Alcaeus:

> Heaven gave this Lyre, – and thus decreed
> Be thou a *bruised*, but not a *broken* reed.

lair's *The Grave* had sparked off the poet in his schooldays nd he wrote his own poem under this title in 1804. Iontgomery continued to be a keen observer of flora and auna. *The Glow-worm* is prefaced: *The male of this insect said to be a fly, which the female caterpillar attracts in he night by the lustre of her train.* Another title is: *A Field 'lower: on finding One in full bloom on Christmas Day.*

n 1805, he was, to quote his biographer, J. W. King, *almost 1 the lion's jaws, for lashing the cowardice of General Mack, vho, with 39,000 Austrians, gave up fighting and bravery nder the walls of Ulm, to the glory and great satisfaction of 'onaparte's military campaign in Germany.* The same year e also published a tract for a wealthy and respectable nember of the Society of Friends with the controversial itle *A Soldier No Christian.* Perhaps he was saved from rosecution by a patriotic editorial on reporting the Battle f Trafalgar.

> Unhappy he whose hopeless eye
> Turns to the light of love in vain;
> Whose cynosure is in the sky,
> He on the dark and lonely main.
>
> *James Montgomery: From 'The Glow-worm'*

> No! – they have 'scaped the waves,
> 'Scaped the sea-monsters' maws;
> They come! but O, shall *Gallic Slaves*
> Give *English Freemen* laws?
>
> *James Montgomery: From 'Ode to the Volunteers of Britain, on the Prospect of Invasion'*

> It smiles upon the lap of May,
> To sultry August spreads its charms,
> Lights pale October on his way,
> And twines December's arms.
>
> . . . . .
>
> On waste and woodland, rock and plain,
> Its humble buds unheeded rise;
> The Rose has but a summer-reign,
> The *Daisy* never dies.
>
> *James Montgomery: From 'A Field Flower: on finding One in full Bloom, on Christmas Day, 1803*

The rumours of Invasion have increased so much in number and assurance, during the past week, that he must be a credulous infidel indeed who can believe that they are utterly ungrounded: but like a structure of mud on a foundation of marble, though falsehood may be founded on truth, the building derives no credit from the base, the base no dishonour from the building; the one is marble, the other mud; the one, truth, the other falsehood; and each as distinct as if they were separated to the distance of the poles.

Editorial comment in 'The Iris' of 13th October 1803

I am concerned to hear that you have lately been so much indisposed; your body is never in perfect tune, – nor is my soul. I have been thinking that between us we might make two very different men, the one the happiest, the other the most miserable in existence.

From a letter of 21st March 1804 from James Montgomery to Joseph Aston.

As a prodigious quantity of counterfeit silver is at present in circulation, we have been advised to recommend to our readers the use of a piece of plain unpolished Italian marble for the proof of half-crowns, shillings and sixpences. The good when passed edgeway on the face of the marble go roughly and hardly along; the bad slide smoothly and easily over it.

From 'The Iris' of 22nd March 1804

We are informed that a number of disorderly boys frequent our market Place, particularly on Saturday nights, which in the general harry and throng, which they increase as much as possible, by pressing and crowding together, they pick the pockets, and pilfer the baskets of unguarded people.
'The Iris' reporting on 28th February 1805 about life in Sheffield

An early communication on some subject connected with the object of the Institution will be esteemed a favour and hope you will regularly comply with the 12th Rule *That all corresponding members shall furnish the Society with a Paper at least every six months.*
From J. Barber, Manchester Philological Society to James Montgomery, 12th September 1805

From the curious and striking changes which were produced upon Matter, and which must be considered as specimens of others equally curious and striking, we do not wonder that Chemistry was once the darling passion of the avaricious, and the phantom which deluded the hopes of the visionary. For the honour of the present age however, it is now patronised by men of science and enlightened judgment, and is brought to such perfection as to gain a most respectable place among the arts. In the clearness of its principles, the solidity of its conclusions, its reference to economical and philosophical uses, it is inferior to none.
Editorial comment in 'The Iris' of 6th March 1806, following a series of lectures on 'Experimental Chemistry' given by Dr. Stancliffe.

His friend Charles Sylvester, described by diarist Thomas Asline Ward as a very entertaining companion, spoke to the Society for the Promotion of Useful Knowledge to which Montgomery belonged on Galvanism, named after Luigi Galvani, who first described the phenomenon of electricity developed by chemical action in 1792. The former plated-wire worker now made most of his living lecturing in this field. Montgomery's poetry included *The Common Lot*, a birthday meditation during a seven mile walk between a Derbyshire village and Sheffield, when the ground was covered with snow, the sky serene, and the mountain air intensely pure. He also published *The Thunderstorm* written not long after witnessing a storm at Scarborough in which three persons were killed by lightning, and assisted another poet Barbara Hoole, later Hofland after marrying the landscape painter Thomas Christopher Hofland, by publishing her poems and assisting in her election as an Associate of the Liverpool Academy of Arts. In his lifetime Montgomery gave encouragement to numerous minor poets. The same year, Montgomery was modelled by the youthful Francis Chantrey, famous later for his *Sleeping Children* in Lichfield Cathedral. William Paulet Carey, an art critic was to encourage the sculptor Chantrey and the poet Montgomery by helping their work to be widely known.

In 1806, Montgomery wrote *Pope's Willow*, verses written for an urn made out of weeping willow, imported from the East, and planted by Pope in his grounds at Twickenham where it flourished for many years but, falling into decay lately broke down.

. . . to kindle in that hallow'd urn
A flame that would for ever burn.

Over all the years of Montgomery's editorship, the paper gave little space to local news. It was primarily concerned with national affairs, giving many full reports of Parliamentary sittings. In 1807 the business was flourishing sufficiently for him to take on an extra compositor and competition from a new rival the *Sheffield Mercury* did not affect circulation to any significant extent. That year, *a local association for the purpose of superseding the employment of Chimney Boys in sweeping chimneys and bettering the conditions of those who were already so engaged* was formed. After several years, it did manage to see a Bill introduced into Parliament but it failed. *The Climbing Boys* part of which was to be read at his funeral service, was written in 1807 to further the cause. The Society did manage to provide an annual lunch each Easter Monday for the lads. That same year, he also wrote

Oh! in this fond susceptive breast,
Dread *Apathy!* thou awful Guest!
Erect thine ebon throne;
Touch'd by the wand of chilling steel,
This throbbing heart shall cease to feel,
This tongue forget to moan.

Barbara Hoole: 'Ode to Apathy'

God of Vengeance, from above
While thine awful bolts are hurl'd,
O remember thou art Love!
Spare! O spare a guilty world!
Stay Thy flaming wrath awhile,
See Thy bow of promise smile.

James Montgomery: From 'The Thunder-Storm'

No woman that has a mother's heart can endure the thought of her own infant being pressed into so abominable a service, and no man that has a father's spirit would permit his chimnies to be thus cleaned, if any other means could be found that would answer the purpose. Such other means are found, and if the public will only give encouragement, there cannot be a doubt that improved machines will completely obviate the necessity of employing climbing boys, when those who use them become expert and expeditious in their business.

. . . . .

One day last week, a Chimney Sweeper's Boy, in attending a Chimney, in the neighbourhood of Carver Street, wedged himself so fast that he could neither get forward nor backward. His cries in this situation were dreadful, and he remained thus pinioned for about an hour, till the Chimney was broken open from the outside of the wall to let him at liberty!

From 'The Iris' of 25th August 1807

Hard fare, cold lodgings, cruel toil,
Youth, health, and strength consume:
What tree could thrive in such a soil?
What flower so scathed could bloom?

Should I outgrow this crippling work,
How should my bread be sought?
Must I to other lads turn Turk,
And teach what I am taught?

. . . . .

For out he glides in April showers,
Lies snug when storms prevail;
He feeds on fruit, he sleeps on flowers –
I wish I was a snail!

James Montgomery: Verses from 'The Complaint' one of 'The Climbing Boy's Soliloquies'

Thus lived the Negro in his native land,
Till Christian cruisers anchor'd on his strand:
Where'er their grasping arms the spoilers spread,
The Negro joys, the Negro's virtues, fled;
Till, far amidst the wilderness unknown,
They flourish'd in the sight of Heaven alone:
While from the coast, with wide and wider sweep,
The race of Mammon dragg'd across the deep
Their sable victims, to that western bourne,
From which no traveller might e'er return,
To blazen in the ears of future slaves
The secrets of the world beyond the waves.

James Montgomery: From 'The West Indies'

Nay, James Montgomery vouched for the fact that, in Sheffield, young girls even were employed in this lung-destroying task. In one case, a girl was set to the work, dressed in boy's clothes.

From the diary of Thomas Asline Ward, 6th May 1807, published in 'Peeps from the Past'

. . . the idea of giving to the world the sacred compositions of Haydn, Beethoven & Mozart with English words but hitherto I have been incapable of perfecting my plan on account of there being no Hymns in the language of the proper measure . . .

The composer William Gardiner writes to James Montgomery on 21st October 1807. William Gardiner's 'Sacred Melodies' was published in 1812

The West Indies in honour of the Abolition of the Slave Trade in the British Legislature, a subject dear to the heart of the son of two missionaries there. This national triumph was perhaps some compensation for the personal despair and disappointment from bad notices given by the Edinburgh reviewers to his recent poetry. He submitted his new poetry to local friends for comment before publication; and John Bailey in particular, wrote his mind. He also wrote critically of William Cobbett's *Vindication of War*. Montgomery was a man with peace in his heart.

The diary of Thomas Asline Ward, a contributor to the *Iris*, contains many references to Montgomery, varying in 1808 from tea taken at Norton Bowling Green to a meeting held in Cutlers' Hall congratulating the King *on the signal and successful efforts of Spain against the perfidy and tyranny of France*. The initiative for the meeting was in an editorial in the *Iris*. In 1809 we find Ward and Montgomery active in ensuring tickets are made available for the relief of the poor. Ward refers also to a reading of Montgomery's new poem *The World before the Flood*. Whilst Montgomery was not himself a member of the Book Club, his friends frequently consulted him for his bibliographical assistance. It is not clear whether Montgomery was an actual member or just a frequent guest at the Utile Dulci Club, whose members in turn, provided a frugal dinner at their respective houses. Mr. W. Bowyer of the Historic Gallery, Pall Mall, proposed to publish a tribute of the fine arts in honour of the abolition of the slave trade, and sought a contribution from Montgomery. Montgomery also wrote stanzas himself in the tragic circumstances of a minister being drowned just after his appointment to his first pastorate. For many years Montgomery, with his friends Ebenezer Rhodes and Edward Nanson had spent almost every evening at the Bull Inn. At length, Montgomery felt the practice had degenerated into a habit which should be broken and the nightly gathering ceased in 1811.

There perhaps never was a time when the Manufacturers of Sheffield were in a state of distress, equal to that in which they find themselves at present, by the almost total stagnation of their trade. Men may differ both concerning the causes of this unexampled suffering amongst the most useful class of society, and concerning the best political measures which ought to be taken for its removal; but we should imagine that every one amongst us, who deserves the name of man, would cordially co-operate with others, of every party and persuasion, in acknowledging and performing the sacred duty, which at this time devolves upon all who can afford it, of relieving by voluntary beneficence, as well as by mere Parish Charity, the wants of our unhappy fellow townspeople.

Editorial comment in 'The Iris' of 29th March 1808

Revolving his mysterious lot,
I mourn him, but I praise him not;
Glory to God be given,
Who sent him, like the radiant bow,
His covenant of peace to show;
Athwart the breaking storm to glow,
Then vanish into heaven.

O Church! to whom that youth was dear,
The Angel of thy mercies here,
Behold the path he trod,
A milky way through midnight skies!
– Behold the grave in which he lies;
Even from this dust thy prophet cries,
Prepare to meet thy God.

James Montgomery: From 'Stanzas to the memory of Rev. Thomas Spencer, of Liverpool'

In short this page is altogether too *bloody* and in my humble opinion the worst that ever came from your pen . . .

John Bailey criticises the draft of a poem by Montgomery in a letter of 25th February 1808

To Thy Temple I repair,
Lord, I love to worship there,
When within the veil I meet
Christ before the Mercy-seat.

Thou, through Him art reconciled
I through Him become thy child;
Abba! Father! give me grace,
In Thy courts to seek Thy face.

While Thy glorious praise is sung,
Touch my lips, unloose my tongue,
That my joyful soul may bless
Thee, the Lord my Righteousness.

While the prayers of saints ascend,
God of love, to mine attend;
Hear me, for Thy Spirit pleads,
Hear, for Jesus intercedes.

While I hearken to Thy Law,
Fill my soul with humble awe,
Till Thy Gospel bring to me
Life and immortality.

While Thy ministers proclaim
Peace and pardon in Thy name,
Through their voice, by faith, may I
Hear Thee speaking from the sky.

From Thine House, when I return,
May my heart within me burn,
And at evening let me say,
I have walk'd with God to-day.

James Montgomery: 'A Day in the Lord's Courts' in the version from 'Original Hymns for Christian Worship'

The 10th couplet very much inferior to that in the original copy which is so beautiful that I should wish it to be restored.

From a letter about a further draft of the poem sent to James Montgomery by John Bailey on 25th February 1808

We recommend to the patronage of our benevolent Readers the Sunday School, just opened in Spring Street, where Poor Children are taught *reading* and *writing* gratuitously. Such Institutions at once display and extend the Spirit of the Gospel, and the liberality of the age. Some conscientious persons have objected to the instruction of the Poor in *writing* at Sunday Schools; but if teaching them to read be satisfying the Sabbath of the Lord, teaching them to write must be conferring yet doubly higher honour on that holy day, by making it doubly beneficial to them . . . Our Saviour, on the Sabbath day, not only opened the eyes of him that was born blind, but healed also the man who had a withered hand. – Is not teaching the ignorant to read, giving sight to the blind eye? Is not teaching them to write transfusing virtue through the withered hand?

Editorial comment in 'The Iris' of 23rd May 1809

Ever since the year 1732, the renewed Church of the Brethren have endeavoured to extend the benefits of Christianity to distant heathen nations. From very small beginnings, the Missions established by them, in different parts of the world, have increased to upwards of thirty settlements, in which, about 150 Missionaries are employed, a number scarcely sufficient for the care of about 24,000 converts, from among various heathen tribes . . . The most flourishing Missions at present, are those in Greenland, Labrador, Antigua, St. Kitt's, the Danish West India Islands, and the Cape of Good Hope . . .

From 'A Concise Account of the Present State of the Mission of the United Brethren; (commonly called Moravians)', January 1811

Jesus! our best beloved Friend,
Draw our souls in sweet desire!
Jesus! in love to us descend,
Baptise us with thy Spirit's fire.

On thy redeeming name we call,
Poor and unworthy though we be,
Pardon and sanctify us all,
Let each thy full salvation see.

Our souls and bodies we resign,
To fear and follow thy commands;
O take our hearts – our hearts are thine,
Accept the service of our hands.

Firm, faithful, watching unto prayer,
Our Master's voice will we obey,
Toil in thy vineyard here, and bear
The heat and burden of our day.

Yet Lord! for us a resting place,
In heaven – at thy right hand prepare,
And till we see thee face to face,
Be all our conversation there.

James Montgomery: 'Social Dedication to God' as printed in Collyer's Hymn Book

Twelve hundred chimnies having been swept with the machine, by one man, during the first twelve months, is a convincing proof that the measure is, in a great degree, practicable.

From a report in 'The Iris' of 23rd January 1812 headed 'Chimney Sweeping and the Machine'

There, with new sight, may they behold
Thy counsels, since the world began,
Like morning's gradual beams, unfold
The wonders of Thy love to man.

James Montgomery: From 'Lord! are there eyes that see the sun'

Singing is an *act* of worship; but sitting is not a *posture* of worship. In heaven *prostration* is used – surely on earth, less than *rising* cannot be deemed due reverence.

William Bengo Collyer: From the preface to 'Hymns, partly collected and partly original, designed as a Supplement to Dr. Watts' Psalms and Hymns'

In 1812, W. B. Collyer published his *Hymns, partly collecte and partly original, designed as Supplement to Dr. Watts Psalms and Hymns* and included several of Montgomery' early hymns including *To Thy temple I repair* which ha been sung to LÜBECK (GOTT SEI DANK) and HART (HARTFORD) and *Jesus our best-beloved friend* sun in recent years to WHITEHALL (SANDYS PSALM 8) an ST POLYCARP. The same year Montgomery was involved i the inauguration of the Bible Association in Sheffield. I very different vein he was present when Dr. David Davie demonstrated his invention of a mode of propelling vessel which will not need horses on the shore path but thos present seemed sceptical. In 1813, he encouraged Miss Shor when she tried to launch a Manuscript Magazine and when i failed, gave encouragement to another's attempt calle

James Montgomery

'he *Portfolio* the following year. In 1813, he wrote for the :hristian Adult Schools in Bristol *Lord are these eyes that* *:e the sun.* He showed however a weariness for current ffairs writing from Scarborough to Thomas Ward that *my* *ιind is utterly barren, for it is indeed unsown with the seeds* *f politics; since it was late fallow . . .*

ince breaking with the Moravians, Montgomery had o settled membership of a religious sect. His first friends ı Sheffield were mostly Unitarians, but he also knew lethodists and other Independents and worshipped variously ver the years. He was however dissatisfied with his position nd in 1814, he was re-admitted to the Moravian Church. ‹s there was no Moravian congregation in Sheffield, he sually worshipped with the Methodists for some years but ept in touch with several denominations. Later in life, e was usually to be found worshipping in the Anglican :hurch. In 1815, he preached his only sermons. Whilst taying at a Moravian household in Cheshire, he was called pon *to preach to us and our neighbours*; and, after some asty preparation did so on Psalm 116 and on the Beatitudes. Ie began several years of active help in the running of heffield Infirmary, sometimes serving on the Quarterly oard; and in other years, looking after detail as a member f the small Weekly Board.

On the 12th, July, 1815, Napoleon Bonaparte sailed out of Rochfort harbour, in a small brig, and delivered himself up to a British Officer who received him on board his flag-ship as a Prisoner! **We looked at these words when we had written them, — we looked again and again; — at first they seemed a delusion, — they must have expressed only a thought which had passed through our mind and was gone into non-entity for ever, — a rumour which we had heard among thousands equally incredible, — a conjecture which we had copied from those papers which abound with little else than conjectures; — yet the hand writing remained the same, — the words would not change their places, — and by no twisting or turning or torturing of the sentence would it confess any other meaning than that which was obvious on the face of it, and which we were at length compelled to believe to be the truth, — namely, that the Spoiler and Oppressor of the Continent, after all his exploits and all his crimes, after all his triumphs and all his failures, had thrown himself upon the mercy of our country, — the country which he had most endeavoured to injure, — the country which he had least succeeded in injuring, during his whole career of ambition and vengeance.**

Editorial comment in 'The Iris' 25th July 1815

Resolved that from the Difficulty of obtaining an occasional Night Nurse a third Nurse be permanently engaged to act as a Night Nurse and when not employed as such to assist the other Nurses or make herself useful as a Servant.

From the Minutes of the Quarterly Board of Sheffield Infirmary of which James Montgomery was a member, 22nd December 1815

This is the *sting* of the Tax, and like the *sting of death* it is *sin* – sin in its essence, sin in its influence; – sin in its operation, by wringing more than his share of the average paid by the whole, from the upright man, and sin in its effect by corrupting those who are negatively, weakly, or even tolerably honest, and drawing them into a snare of *falsehood*, or to give it the mildest appellation, *duplicity*. It may be said, with supercilious contempt of what is due to human frailty, in the hour of severe trial, that the Tax does not force people to make lying returns of their profits. No – but it tempts them to do so; – Satan himself cannot force men to sin; but he tempts them to it, and he tempts them successfully, – *so* does the Income Tax. Let him who is without sin amongst us here cast the first stone at another.

From an editorial on the renewal of the Income Tax in 'The Iris' of 19th March 1816

The most important annual event in human affairs is the Meeting of the British Parliament.

Editorial comment in 'The Iris' of 18th February 1817

Command thy blessing from above,
O God! on all assembled here;
Behold us with a father's love,
While we look up with filial fear.

Command thy blessing, Jesus! Lord!
May we thy true disciples be;
Speak to each heart the mighty Word,
Say to the weakest, *Follow me.*

Command thy blessing in this hour,
Spirit of Truth! and fill this place
With humbling and with healing power,
With killing and with quickening grace.

O Thou, our Maker, Saviour, Guide,
One true eternal God confest!
Whom thou hast join'd let none divide,
None dare to curse whom thou hast blest.

With thee and these for ever found,
May all the souls who here unite,
With harps and songs thy throne surround,
Rest in thy love and reign in light.

James Montgomery: 'For a solemn Assembly'

Sinners, wrung with true repentance,
Doom'd for guilt to endless pains,
Justice now revokes the sentence,
Mercy calls you – break your chains:
Come and worship,
Worship Christ the new-born King.

James Montgomery: The final verse of 'Good Tidings of Great Joy to all People' which begins 'Angels from the realms of glory' from 'Original Hymns for Christian Worship'

Bristol! to thee the eye of Albion turns . . .

. . . Health to the sick, and to the hungry bread,
Beneficence to all, their hands shall deal,
With Reynolds' single eye and hallow'd zeal.
Pain, want, misfortune, thither shall repair;
Folly and vice reclaim'd shall worship there
The God of *him* – in whose transcendent mind
Should *such* a temple, free to all mankind:
Thy God, thrice-honour'd city! bids thee raise
That fallen temple, to the end of days:
Obey his voice; fulfil thine high intent;
– Yea, be thyself the *Good Man's Monument!*

James Montgomery: From 'A Good Man's Monument'

On 3rd June 1816, the Whitsuntide gathering of the Sheffield Sunday School Union first sang *Command thy blessing from above*. It was published in the *Evangelical Magazine* that September. When he published it with the definitive texts of his hymns in *Original Hymns for Christian Worship*, it was titled *For a solemn Assembly*. Its tunes include BOW BRICKHILL, NIAGARA, BROCKHAM (CONFIDENCE (Clarke); GILLINGHAM (Clarke)) and EISENACH (LEIPZIG (Schein); MACH'S MIT MER, GOTT; SCHEIN). For the Christmas Eve issue of the *Iris*, he printed the text *Angels from the realms of glory*. It appeared in 1825 in the *Christmas Box* as one of *Three New Carols* in the first complete book published by the *Religious Tract Society*. Its many tunes include IRIS (SHEPHERDS IN THE FIELDS), LEWES, GRAFTON (TANTUM ERGO SACRAMENTUM; TANTUM ERGO (French Melody)), KENSINGTON NEW, WOODFORD GREEN and FENTON COURT.

He also wrote a piece addressed to the members of the Reynolds Commemoration Society. Richard Reynolds was a member of the Society of Friends and one of those who are Christians not only in word but in deed.

Samuel Roberts was another of Montgomery's close friends. In 1817 he wrote *The State Lottery* which included a prophetic speech spaced within a dream of an address delivered to both Houses of Parliament a century later in 1917. Roberts did not anticipate the first world war but did perhaps anticipate the lottery of premium bonds a few decades later. Montgomery wrote a matching poem. This represented a turnabout by Montgomery who, until 1816, had accepted advertisements in the *Iris* for the State Lottery. The year had begun with six distinguished citizens, including Montgomery and Roberts, guaranteeing £150 to the Society for Bettering the Conditions of the Poor in addition to £200 guaranteed earlier. Montgomery also took a former apprentice into partnership at the *Iris*, but the alliance proved disastrous to the gentle poet-editor and it was dissolved the following year. The year also saw Montgomery, through the influence of George Bennett, taking an active part in ensuring the success of Rotherham Independent College previously set up for the purpose of training Congregationalist Ministers.

Life is the most precious of earthly blessings, and liberty is the next to it. Without the former no other can be possessed; without the latter none can be enjoyed in security. War, therefore, the determined destroyer of life, must be the greatest crime that Society can commit, and the enslavement of human beings is second only in atrocity to that.

Editorial comment in 'The Iris' of 27th January 1818

Hark! the song of jubilee was composed at the express desire of the London Missionary Society with a special reference to the renunciation of idolatry and acknowledgement of the Gospel in the Georgian Isles of the

Lord, teach us how to pray aright,
With reverence and with fear;
Though dust and ashes in Thy sight,
We may, we must draw near.

We perish if we cease from prayer;
Oh! grant us power to pray;
And when to meet Thee we prepare,
Lord, meet us by the way.

Burden'd with guilt, convinced of sin,
In weakness, want, and woe,
Fightings without, and fears within,
Lord, whither shall we go?

God of all grace, we bring to Thee
A broken, contrite heart;
Give, what Thine eye delights to see,
Truth in the inward part.

Give deep humility; the sense
Of godly sorrow give;
A strong, desiring confidence
To hear Thy voice and live; –

Faith in the only Sacrifice
That can for sin atone;
To cast our hopes, to fix our eyes,
On Christ, on Christ alone; –

Patience to watch, and wait, and weep
Though mercy long delay;
Courage, our fainting souls to keep,
And trust Thee though Thou slay.

Give these, and then Thy will be done;
Thus, strengthen'd with all might,
We, through Thy Spirit and Thy Son,
Shall pray, and pray aright.

James Montgomery: 'The Preparation of the Heart'
This version is from 'Original Hymns for Christian
Worship' as slightly revised by the author

Prayer is the soul's sincere desire,
Utter'd or unexpress'd,
The motion of a hidden fire
That trembles in the breast.

Prayer is the burthen of a sigh,
The falling of a tear,
The upward glancing of an eye,
When none but God is near.

Prayer is the simplest form of speech
That infant-lips can try,
Prayer the sublimest strains that reach
The Majesty on high.

Prayer is the contrite sinner's voice,
Returning from his ways,
While angels in their songs rejoice,
And cry, *Behold, he prays!*

Prayer is the Christian's vital breath,
The Christian's native air,
His watchword at the gates of death;
He enters heaven with prayer.

The saints in prayer appear as one
In word, and deed, and mind,
While with the Father and the Son
Sweet fellowship they find.

Nor prayer is made by man alone,
The Holy Spirit pleads,
And Jesus, on the eternal throne,
For sinners intercedes.

O Thou by whom we come to God,
The life, the truth, the way!
The path of prayer Thyself hast trod:
Lord, teach us how to pray.

James Montgomery: 'What is Prayer?' in his revised
version from 'Original Hymns for Christian Worship'

Southern Seas and was sung on 14th May 1818 and published in the *Evangelical Magazine* that July. The singers may have been tongue-tied by his original use of *abysses* for *depths* in verse 2. Tunes have included THANKSGIVING (Gilbert) and ST GEORGE'S, WINDSOR (ST GEORGE (Elvey)). Another broadsheet that year was published for use in Sheffield's Nonconformist Sunday Schools. *Lord, teach us how to pray aright* has been sung to FIRST MODE MELODY, WINDSOR (DUNDEE (Damon)), ST PETER, ST FRANCES, ST HUGH (Hopkins), MAISEMORE, FARRANT, REGENT SQUARE and BURFORD. There appeared on the same broadsheet *Prayer is the soul's sincere desire* actually written for Edward Bickersteth's *Treatise on Prayer*. It was rejected by some hymn book editors as being inappropriate because it was not addressed to God until the last verse, the earlier stanzas being a recital of facts. Some overcame this objection by repeating the last verse as the first; so the hymn is also known as *O thou by whom we come to God*. Tunes used include WIGTON (WIGTOWN), SONG 67 (PALATINE; ST MATTHIAS (Gibbons)), NOX PRAECESSIT and ROCHESTER.

> Hark! the song of Jubilee;
> Loud as mighty thunders roar,
> Or the fulness of the sea,
> When it breaks upon the shore:
> Hallelujah! for the Lord
> God Omnipotent, shall reign;
> Hallelujah! let the word
> Echo round the earth and main.
>
> Hallelujah! – hark! the sound
> From the abysses to the skies,
> Wakes above, beneath, around,
> All creation's harmonies;
> See Jehovah's banner furl'd,
> Sheath'd His sword: He speaks – 'tis done,
> And the Kingdoms of this world
> Are the Kingdoms of His Son.
>
> He shall reign from pole to pole
> With illimitable sway;
> He shall reign, when like a scroll
> Yonder heavens have pass'd away:
> Then the end; – beneath His rod,
> Man's last enemy shall fall;
> Hallelujah! Christ in God,
> God in Christ, is all in all.
>
> *James Montgomery: 'Hallelujah'*

The Reverend Thomas Cotterill was one of the first Anglican priests to advocate the use of hymns over and above metrical psalms. After serving at Lane End, Staffordshire, he was appointed the incumbent of St Paul's Church, Sheffield. James Montgomery played a considerable part in the production in 1819 of the much enlarged and famous eighth edition of Cotterill's *Selection of Psalms and Hymns*. His firm printed the books; he contributed nearly sixty of his own hymns and used his wide knowledge of available texts to recommend others for inclusion. *Lord God, the Holy Ghost*, sung sometimes to GILDAS (ST AUGUSTINE), DONCASTER (BETHLEHEM (Wesley)), ANNUNCIATION, VENICE, ST ETHELWALD, WATCHMAN (Leach) and BOWDEN was titled *The Descent of the Spirit* and *Songs of praise the angels sang*, whose tunes have included CULBACH (ACH, WANNE KOMMT), RILEY, LÜBECK (GOTT SEI DANK), LAUDS (Wilson) and NORTHAMPTON, have stood

Songs of praise the angels sang,
Heaven with hallelujahs rang,
When Jehovah's work begun,
When He spake, and it was done.

Songs of praise awoke the morn,
When the Prince of Peace was born;
Songs of praise arose, when He
Captive led captivity.

Heaven and earth must pass away,
Songs of praise shall crown that day;
God will make new heavens, new earth,
Songs of praise shall hail their birth.

And can man alone be dumb,
Till that glorious kingdom come?
No; – the Church delights to raise
Psalms, and hymns, and songs of prais

Saints below, with heart and voice,
Still in songs of praise rejoice;
Learning here, by faith and love,
Songs of praise to sing above.

Borne upon their latest breath,
Songs of praise shall conquer death;
Then, amidst eternal joy,
Songs of praise their powers employ.

James Montgomery as printed i
'Original Hymns for Christian Worshi

Lord God, the Holy Ghost,
In this accepted hour,
As on the day of Pentecost,
Descend in all Thy power;
We meet with one accord
In our appointed place,
And wait the promise of our Lord,
The Spirit of all grace.

Like mighty rushing wind
Upon the waves beneath,
Move with one impulse every mind,
One soul, one feeling breathe:
The young, the old inspire
With wisdom from above;
And give us hearts and tongues of fire
To pray, and praise, and love.

Spirit of light, explore,
And chase our gloom away,
With lustre shining more and more
Unto the perfect day:
Spirit of truth, be Thou
In life and death our guide;
O Spirit of adoption, *now*
May we be sanctified.

James Montgomery: 'The Descent of the Spirit
as printed in 'Original Hymns for Christian Worship

Singing is most properly an act of the congregation. It should therefore be congregational. No tongue should be silent which is able to join. Choirs of singers, separate from the congregation, were introduced in the fourth century, when this part of divine service was greatly neglected. *It was the decay of singing which*

the test of time. Part of Cotterill's congregation rebelled against the use of this hymn book. Hymnody was not yet accepted as proper for use in worship by many Anglicans and an appeal was made against the book to the Diocesan Court. Happily the Archbishop of York intervened as a conciliator and a ninth edition with a substituted selection received his Imprimatur. The way however later editors drew on the original eighth edition confirms what a treasury Montgomery and Cotterill had brought together.

The same year, James Montgomery published *Greenland*. His original plan was that the poems should *embrace the most prominent events in the annals of ancient and modern Greenland:- incidental descriptions of whatever is sublime or picturesque in the seasons and scenery, or peculiar in the superstitions, manners, and character of the nations – with a rapid retrospect of that moral revolution, which the Gospel has wrought among that people, by redeeming them, almost universally, from idolatry and barbarism.* Of the five cantos written, the first three contain a sketch of the history of the ancient Moravian Church, the origin of the missions by the Moravians to Greenland, and the voyage of the first three brethren who went thither in 1733. The fourth canto refers principally to traditions concerning the Norwegian colonies, which were said to have existed on both shores of Greenland, from the tenth to the fifteenth centuries. In the fifth canto, the author has attempted, in a series of episodes, to sum up and exemplify the chief causes of the extinction of those colonies, and the abandonment of Greenland for some centuries by European voyagers.

***first brought this order of singers into the Church*; and their introduction was only meant as a temporary provision. The Psalms are appointed to be sung *of all the people together.* Choirs of singers are, however, very useful, and necessary, when they act, not as sole performers, but as the leaders of congregations. It is a perversion of this holy employment, and derogatory to the honour of God to assign the work of singing his praises to a few individuals.**

T. Cotterill: From the preface to the eighth edition of 'A Selection of Psalms and Hymns for public and private Use adapted to the Services of the Church of England'

The story of the introduction of Christianity among the Sclavonic tribes is interesting. The Bulgarians, being borderers on the Greek empire, frequently made predatory incursions on the Imperial territory. On one occasion the sister of *Bogaris*, King of the Bulgarians, was taken prisoner, and carried to Constantinople. Being a royal captive, she was treated with great honour, and diligently instructed in the doctrines of the Gospel, of the truth of which she became so deeply convinced, that she desired to be baptized; and when, in 845, the Emperor *Michael III* made peace with the Bulgarians, she returned to her country a pious and zealous Christian. Being earnestly concerned for the conversion of her brother and his people, she wrote to Constantinople for teachers to instruct them in the way of righteousness. Two distinguished bishops of the Greek Church, *Cyrillus* and *Methodius*, were accordingly sent into Bulgaria. The King *Bogaris*, who heretofore had resisted conviction, conceived a particular affection for *Methodius*, who, being a skilful painter, was desired by him, in the spirit of a barbarian, to compose a picture exhibiting the most horrible devices. *Methodius* took a happy advantage of this strange request, and painted the day of judgment in a style so terrific, and explained its scenes to his royal master in language so awful and affecting, that *Bogaris* was awakened, made a profession of the true faith, and was baptized by the name of *Michael*, in honour of his benefactor the Greek Emperor. His subjects, according to the fashion of the times, some by choice, and others from constraint, adopted their master's religion. To *Cyrillus* is attributed the translation of the

Yesterday, the Ninth Anniversary Meeting of the Female Friendly Society was held at the Cutlers' Hall, and attended by a numerous assembly of the Ladies who patronise it and conduct it. These heard of twenty-five widows and single women, from 65 up to 108 years of age, have been relieved from its small funds during the past twelve months, at an average of less than sixteen shillings to each; and such is the poverty and misery of these objects of Christian benevolence, that this small sum has been in every instance, we believe, as a great and seasonable relief.

From a report in 'The Iris' of 11th May 1819

Scriptures still in use among the descendants of the Sclavonian tribes, which adhere to the Greek Church; and this is probably the most ancient European version of the Bible in a living tongue.

Most heartily do we congratulate our countrymen on what may be called, without a metaphor, a national deliverance, in the abandonment of the proceedings against the Queen . . . we heartily congratulate Her Majesty on her deliverance from this fiery ideal, as, in the beginning of this article, we have congratulated our countrymen on this national deliverance from the greatest evil, that of civil war, which could befall a people.

Editorial comment in 'The Iris' of 14th November 1820 concerning the divorce proceedings against Queen Caroline

But notwithstanding this triumphant introduction of Christianity among these fierce nations (including the Bohemians and Moravians), multitudes adhered to idolatry, and among the nobles especially many continued Pagans, and in open or secret enmity against the new religion and its professors. In Bohemia, Duke *Borziwog*, having embraced the Gospel, was expelled by his chieftains, and one *Stoymirus*, who had been thirteen years in exile, and who was believed to be a heathen, was chosen by them as their prince. He being, however, soon detected in Christian worship, was deposed, and *Borziwog* recalled. The latter died soon after his restoration, leaving his widow, *Ludomilla*, regent during the minority of her son *Wratislaus*, who married a noble lady, named *Drahomira*. The young duchess, to ingratiate herself with her husband and her mother-in-law, affected to embrace Christianity, while in her heart she remained an implacable enemy to it. Her husband dying early, left her with two infant boys. *Wenceslaus*, the elder, was taken by his grandmother, the pious *Ludomilla*, and carefully educated in Christian principles; the younger, *Boleslas*, was not less carefully educated in hostility against them by *Drahomira*; who, seizing the government during the minority of her children, shut up the churches, forbade the clergy either to preach or teach in schools, and imprisoned, banished, or put to death those who disobeyed her edicts against the Gospel. But when her eldest son, *Wenceslaus*, became of age, he was persuaded by his grandmother and the principal Christian nobles to take possession of the government, which was his inheritance. He did so, and began his reign by removing his pagan mother and brother to a distance from the metropolis. *Drahomira*, transported with rage, resolved to rid herself of her mother-in-law, whose influence over *Wenceslaus* was predominant. She found two heathen assassins ready for her purpose, who, stealing unperceived into *Ludomilla's* oratory, fell upon her as she entered it for evening prayers, threw a rope round her neck, and strangled her. The remorseless *Drahomira* next plotted against *Wenceslaus*, to deprive him of the government; but her intrigues miscarrying, she proposed to her heathen son to murder him. An opportunity soon offered. On the birth of a son, *Boleslas* invited his Christian brother to visit him, and be present at a pretended ceremony of blessing the infant. *Wenceslaus* attended, and was treated with unwonted kindness; but, suspecting treachery, he could not sleep in his brother's house. He therefore went to spend the night in the church. Here, as he lay defenceless in an imagined sanctuary, *Boleslas*, instigated by their unnatural mother, surprised and slew him with his sabre. The murderer immediately usurped the sovereignty, and

My legs . . . the complaint in them has been so bad of late, and continues so inveterate, that I thought I should not do right if I did not try Harrogate water.

James Montgomery writes to his niece Betsy from Harrogate on 20th October 1820

commenced a cruel persecution against the Christians, which was terminated by the interference of the Roman Emperor *Otto I*, who made war upon *Boleslas*, reduced him to the condition of a vassal, and gave peace to his persecuted subjects. This happened in the year 943.

James Montgomery: From 'Appendix to Greenland'

The circus exhibitions of Messrs. Adams and Cooke, and the wild animal ones of Drake and Shore, and Atkin, displayed an appearance of grandeur and emulation that seemed to afford pleasure to the spectators, who might be said to be a countless multitude of respectable-looking persons of all ages and both sexes . . . Mr. Shore, a native of this town, very kindly permitted the Boys and Girls of the Charity School, and the Girls of the National School, to view his superior collection of wild animals, on Friday afternoon.

From a report in 'The Iris' 1821

How speed the faithful witnesses, who bore
The Bible and its hopes to Greenland's shore?
– Like Noah's ark, alone upon the wave
(Of one lost world the 'immeasurable grave),
Yonder the ship, a solitary speck,
Comes bounding from the horizon; while on deck
Again imagination rests her wing,
And smooths her pinions, while the Pilgrims sing
Their vesper oraisons. – The Sun retires,
Not as he wont, with clear and golden fires;
Bewilder'd in a labyrinth of haze,
His orb redoubled, with discolour'd rays,
Struggles and vanishes; – along the deep,
With slow array, expanding vapours creep,
Whose folds, in twilight's yellow glare uncurl'd,
Present the dreams of an unreal world;
Islands in air suspended; marching ghosts
Of armies, shapes of castles, winding coasts,
Navies at anchor, mountains, woods, and streams,
Where all is strange, and nothing what it seems;
Till deep involving gloom, without a spark
Of star, moon, meteor, desolately dark,
Seals up the vision: – then, the Pilot's fears
Slacken his arm; a doubtful course he steers,
Till morning comes, but comes not clad in light;
Uprisen day is but a paler night,
Revealing not a glimpse of sea or sky;
The ship's circumference bounds the sailor's eye.
So cold and dense the impervious fog extends,
He might have touch'd the point where being ends;
His bark is all the universe; so void
The scene, – as though creation were destroy'd,
And he and his few mates, of all their race,
Were here becalm'd in everlasting space.

James Montgomery: From 'Greenland'

Fair Moon, why does thou wane?
– That I may wax again.

James Montgomery: From 'Questions and Answers'

Beside tne objections, which grow ever weightier on reflection, against the effect of exposing persons in the service of religion but not under its influence to the satire which they deserve for themselves, though it can scarcely be inflicted on them except at the expense of *what* is sacred . . .

From a letter of 24th January 1821 from James Montgomery to James Everett

Calvary's mournful mountain climb,
There, adoring at His feet,
Mark that miracle of time,
God's own sacrifice complete:
It is finish'd; – hear the cry;
Learn of Jesus Christ to die.

James Montgomery: Verse 3 of 'Christ our Example in Suffering' which begins 'Go to dark Gethsemane' in the final version in 'Original Hymns for Christian Worship'

By such shall He be feared,
While sun and moon endure,
Beloved, obey'd, revered;
For He shall judge the poor,
Through changing generations,
With justice, mercy, truth,
While stars maintain their stations,
Or moons renew their youth.

James Montgomery: The third verse of 'The Reign of Christ on Earth' which begins 'Hail to the Lord's Anointed!' as printed in 'Original Hymns for Christian Worship'

Thrice welcome, little English flower!
My mother-country's white and red,
In rose or lily, till this hour,
Never to me such beauty spread:
Transplanted from thine island-bed,
A treasure in a grain of earth,
Strange as a spirit from the dead,
Thine embryo sprang to birth.

James Montgomery: Opening verse of 'The Daisy in India'

Montgomery wrote *Go to dar Gethsemane* for Cotterill's selection fo 1820 but rewrote it later for *Th Christian Psalmist*. It is sung to NICH SO TRAURIG (Bach), LLYFNANT PRESSBURG (NICHT SO TRAURI (Freylinghausen)), GETHSEMAN (Monk), and PETRA (REDHEAI No.76; AJALON; NORWOOD (SA)) Montgomery was one of a group o six citizens who met to discuss th distress of the traders in Sheffield. I consequence a subscription was raised t assist workmen who could not obtai appointments at the regular rate.

George Bennett, who was a close frien of Montgomery for many years, agree to visit overseas stations for the Londo Missionary Society. James wrote o 10th March, 1821 *Verses to my Friend George Bennett Esq., of Sheffield, o his intended Voyage to Tahiti, and othe Islands of the South Sea, wher Christianity has recently bee established.*

Hail to the Lord's Anointed is one o Montgomery's best loved texts, sun over the years to CRÜGE (HERRNHÜT (Crüger)), LANCASHIR (Smart), HOLMBRIDGE, MUNIC (BREMEN) and HOLY CHURCH. Base on Psalm 72, it was sung that Christma at the Moravian Settlement at Fulneck and sent the followin month to a missionary in the South Seas. But its break through came when Dr. Adam Clarke, who had been th chairman of a missionary meeting at which Montgomer recited it, printed it with a special note in his renowne Commentary on the whole Books of Scripture. Montgomer found himself in difficulty with John Everett for refusing t print material which might offend. Everett was eventuall expelled by the Wesleyan Conference for publishing libel against fellow Wesleyans.

I know not that I ever enjoyed, since leaving Europe, a simple pleasure so exquisite as the sight of this *English* Daisy afforded me; not having seen one for upwards of thirty years, and never expecting to see one again.

William Carey, in a letter acknowledging receipt of seeds from England

In 1822, Montgomery published *Songs of Zion, being Imitations of Psalms.* The sixty-seven items included *O God, Thou art my God alone*, based on Psalm 63, sung sometimes to WAINWRIGHT and *God is my strong salvation*, based on

Psalm 27, sung to CHRISTUS DER IST MEIN LEBEN (BREMEN; PASTOR; VULPIUS). Amongst his poetry is *The Daisy in India.* A friend, a scientific botanist, sent a packet of sundry seeds to Dr. William Carey, the Baptist missionary working in India which had been gratefully acknowledged. Montgomery was a considerable correspondent and amongst those to whom he wrote was Joseph Aston another journalist, whose *A Metrical History of Manchester* was published that year.

In 1823 Montgomery was appointed one of the four Vice-Presidents of the newly constituted Sheffield Philosophical and Literary Society. This Society eventually assembled much of Montgomery's correspondence for posterity and hundreds of letters show the width of his circle.

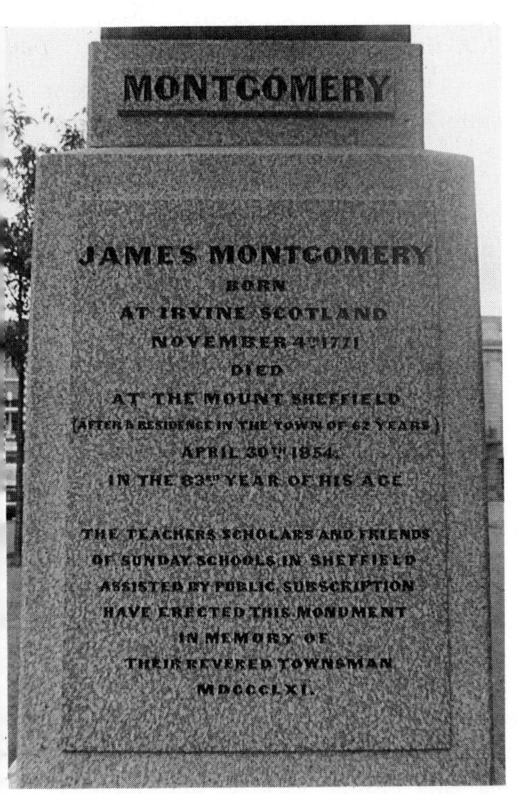

Inscription at base of memorial, Sheffield

Launch boldly on the surge,
And in a light and fragile bark
Thy path through flood and tempest urge,
Like Noah in the ark;
Then tread like him, a new world's shore,
Thine altar build, and God adore.

. . . . .

While these enchant thine eye
O, think how often we have walk'd,
Gazed on the glories of *our* sky,
Of higher glories talk'd,
Till our hearts caught the kindling ray,
And burn'd within us by the way.

James Montgomery: From 'To my friend, George Bennett, Esq.

In Seventeen Hundred and Eighty and Seven,
The Church of *St James* was devoted to Heaven,
Where to offer up prayer, and learn Divine Will,
Christians met; – whilst, for wicked men who do ill,
The Foundation Stone of the *New Bayley Prison*,
By the Hundred was laid, – and, for grim Law's decision,
A *Court-House* appendant – for till then – it was strange! –
The Sessions were held in a Room at the 'Change!

Joseph Aston: From 'A Metrical History of Manchester'

God is my strong salvation,
What foe have I to fear?
In darkness and temptation,
My light, my help is near;
Though hosts encamp around me,
Firm to the fight I stand:
What terror can confound me,
With God at my right hand?

Place on the Lord reliance,
My soul, with courage wait;
His truth be thine affiance,
When faint and desolate:
His might thine heart shall strengthen,
His love thy joy increase;
Mercy thy days shall lengthen,
The Lord will give thee peace.

James Montgomery: 'Trust in the Lord'

A selection from the wide range of James Montgomery's correspondents

George Henry Robbins (1778 - 1847) was an auctioneer and philanthropist. He was on one occasion arrested as a beggar when trying to raise funds for a Sea Bathing Infirmary at Margate, one of the many causes he supported. He urged public sympathy, too, for comedians who had fallen on hard times.

I send you the largest apple that has grown in our garden — I wish it was as good as it is great, but there are persons of our sex who do not think *that* to be necessary in regard to themselves; although they would be the first to require it of an apple.
James Montgomery writing to Sarah Gales whilst she is away in Brighton on 5th September 1822

Richard Ryan (1796 - 1849) was a biographer, bookseller and playwright. He wrote *Le Pauvre Jacques*, a vaudeville in one act from the French and *Poetry and Poets*, being a collection of the choicest anecdotes relative to the poets of every age and place.

John Aikin (1747 - 1822) was a physician and author of *Essays on Song Writing.*

Thomas Foster Barham (1766 - 1844) was a musician and author. He wrote *Abdullah or the Arabian Martyr* and composed *Musical Meditations.*

Josiah Pratt (1768 - 1844) was Secretary of the Church Missionary Society, one of the founders of the British and Foreign Bible Society and publisher of the *Missionary Register.*

John Sheppard (1785 - 1879) was an Anabaptist. This traveller and religious author wrote *Thoughts preparative or persuasive to Private Devotions.*

Bow in the cloud, what token does thou bear?
— That Justice still cries *strike* and Mercy *spare.*
James Montgomery: From 'Questions and Answers'

Elizabeth Anne Le Noir (?1755 - 1841) was poet, novelist and co-owner of the *Reading Mercury*. She wrote *Victorine's Excursions.*

James Ballantyne (1772 - 1833) was printer of the works of Sir Walter Scott and publisher of *Minstrelsey of the Scottish Border.*

John Antes La Trobe (1799 - 1878) was an Anglican priest and author of *The Music of the Church considered in its various Branches, Congregational and Choral.*

Bernard Barton (1784 - 1849) was a Quaker poet who spent his working life with Messrs. Alexander's Bank. His *Convict's Appeal* is a protest in verse against the severe criminal laws of his day.

Frederic Shoberl (1775 - 1853) who was educated at the Fulneck Moravian Settlement, was co-proprietor and editor of *New Monthly Magazine* and later assisted Rudolph Ackermann on *Forget-me-not.*

Rudolph Ackermann (1764 - 1834) born at Stolberg in Saxony, oversaw a fund of £200,000, half received from public donations, for the relief of suffering in Germany after the Battle of Leipzig. This fine-art publisher and bookseller, issued *Forget-me-not*, an annual begun in 1825, unparalleled in typographical and artistic merit.

Charles Smith Bird (1795 - 1862) was an Anglican cleric who wrote *For Ever*. He was an ardent amateur entomologist.

Francis Hodgson (1781 - 1852) was a poet who enjoyed the friendship of Lord Byron. He became Provost of Eton.

Thomas Grinfield (1788 - 1870) was a cleric and minor hymn-writer. He wrote *The Visions of Patmos*.

George Hadfield (1787 - 1879) was an attorney and member of Parliament who took a liberal view of politics. He was active in the Anti Corn Law League.

John Forster (1812 - 1876) was a historian and biographer. *The Forster Collection*, bequeathed by will, consisted of around eighteen thousand books.

William Gardiner (1770 - 1853) worked first in the hosiery trade. He published the *Sacred Melodies from Haydn, Mozart, Beethoven, and other Composers, adapted to the best Poets appropriated for use in the British Church.* The composer enjoyed a reputation as a conversationalist.

John Bishop Estlin (1785 - 1855) was an opthalmic surgeon who established a dispensary for the poor in Frogmore Street, Bristol and personally treated over 52,000 poor patients. He collected additional supplies of vaccine lymph from cows near Berkeley in Gloucestershire for the prevention of smallpox, and campaigned against slavery and medical impostors and in favour of religious tolerance. He was a Unitarian.

John Dix (?1800 - ?1865) the biographer of Thomas Chatterton was thought by many to be a rogue who had made up untrue biographical material.

William Ellis (1794 - 1872) was a missionary in South Africa and in the South Sea Islands. Later he served as Chief Foreign Secretary of the London Missionary Society. He edited an annual *The Christian Keepsake.*

Richard Alfred Davenport (1777 - 1852) was a collector of books, manuscripts, pictures, ancient coins and antiques. He

It is one of the most difficult questions in the administration of justice, how far individuals in a free country like ours may be prosecuted for blasphemous publications. We believe that it will not be denied by the stoutest champion of the liberty of the press, who has any respect for decorum in private life, or good order in civil government, that works grossly obscene, or flagrantly seditious, should be repressed by the penal interference of the law, indeed this is quite as reasonable and necessary as it is that modesty should be protected from personal outrage, or allegiance from overt acts of treason. Wise and good people, however, have differed much on the subject first mentioned, though no man, we fear not to say, can be either wise or good, who would tolerate the diffusion of principles which he believed in his heart were dangerous both to the present and eternal welfare of those who might be seduced to receive them, provided they could be prohibited by less peril than permitted.

From editorial comment in 'The Iris' of 19th October 1822

I have no prospect of forwarding any letters to Europe until we shall ourselves go to the colony of New South Wales and forward them from thence. which voyage we however hope will now be in six weeks or two months. I will however indulge myself in writing you a few lines which will be ready should an opportunity suddenly present itself. An opportunity very unexpectedly occurred about 2 months ago by an English whaler Capt. Brookes.

George Bennet writing to James Montgomery from Erineo on 26th January 1824. Their friendship survived long periods of separation and silence.

wrote *Lives of Individuals who raised Themselves from Poverty to Eminence and Fortune.*

Robert Hartley Cromek (1770 - 1812) was an engraver and book collector who published *Select Scottish Songs, Ancient and Modern* with notes by Robert Burns. Cromek earned a reputation as a shifty speculator making money out of the genius of others.

Samuel Carter Hall (1800 - 1889) was a reviewer, reporter and publisher who founded *The Amulet*, a Christian and Literary Remembrancer. He wrote *The Acquittal of the Seven Bishops*, one of whom was Thomas Ken, the hymnwriter.

James Baldwin Brown (1785 - 1843) was a barrister on the northern circuit, and later a judge. He published *Memoirs of the Public and Private Life of John Howard, the Philanthropist.*

Thomas Easthoe Abbott (1779 - 1854) who spent most of his life in Darlington, was a poet interested in the missionary movement. He wrote *The Triumph of Christianity.*

Thomas Pringle (1789 - 1834) who was lame, worked for many years in the Registry Office as a copyist. He was the first co-editor of the *Edinburgh Monthly Magazine*, and on emigrating to South Africa, had his newspapers the *South African Journal* and the *South African Commercial Advertiser* banned by the Governor. He returned home to face a difficult financial situation.

Henry Neele (1798 - 1828) was a map and heraldic engraver, a poet and short story writer.

Rann Kennedy (1772 - 1851) was second master at St. Edward's School, Birmingham. He edited *A Church of England Psalm-book or portion of the Psalter adapted . . . to the Services of the Church of England.*

Money is literally become merchandize in this country: there is a greater trade carried on in it than in any thing else. Cotton and cotton-spinning, marvellous as the manufacture of the cheap material into immeasurable and inestimable articles appears are insignificant in comparison with the multitude, immensity, and value of forms into which *Plutus*, turned *Proteus*, metamorphoses himself in this age of wealth and speculation.

Editorial comment in 'The Iris' on 26th May 1824

Thomas Jackson (1783 - 1873) twice President of the Wesleyan Conference, edited the works of John Wesley, wrote *The Life of Charles Wesley* and held the Chair of Divinity at Richmond College.

Joseph Hunter (1783 - 1861) was a Presbyterian Minister and Sub-Commissioner of Public Records. Although living away, he was author of *Hallamshire: the History and Topography of Sheffield.*

Thomas Henry Illidge (1799 - 1851) was a portrait painter, many of whose works were hung in the Liverpool Academy and the Royal Academy.

William Henry Lyttleton (1782 - 1837) was a politician who campaigned on behalf of the climbing boys and against state lotteries.

William Pickering (1796 - 1854) published *The Diamond Classics.* His business was renowned for the excellency of its typography and cloth binding.

William Maxwell Hetherington (1803 - 1869) was a Church of Scotland minister and professor of theology. He wrote *Twelve Dramatic Sketches founded on the Pastoral Poetry of Scotland.*

Edward Bickersteth (1786 - 1850) travelled widely for and became Secretary of the Church Missionary Society. He compiled *Christian Psalmody.*

William Paulet Carey (1759 - 1839) was an Irish painter, engraver and art critic. He provided the copper plates in Geoffrey Gambado's (H. Bunbury's) *Annals of Horsemanship.* He recognised early the genius of Francis Chantrey as a sculptor and James Montgomery as a poet.

Josiah Conder (1789 - 1855) was a bookseller and author who compiled *The Congregational Hymn Book* in 1834.

Thomas Raffles (1788 - 1863) was a renowned independent minister in Liverpool and sometime Chairman of the Congregational Union. He was one of the founders of Blackburn Academy to train candidates for the ministry. His use of anecdotes was legion.

Charles Crocker (1797 - 1861) was a poor poet and shoemaker. Later he served as Bishop's verger at Chichester Cathedral.

George Daniel (1789 - 1864) was known as a lover of old books, old wines, old customs and old friends. Of Huguenot descent, he wrote the musical farce *The Disagreeable Surprise* and collected theatrical curiosities.

Charles Cowden Clarke (1787 - 1877) became a partner of the music publisher, Albert Novello. He loved the theatre as well as music, cricket as well as fine arts and lectured widely.

James Everett (1784 - 1872) was a practical joker before becoming a bookseller and a Wesleyan Minister. He was expelled at the Conference in 1849 and became first President of the Assembly of the breakaway United Methodist Free Church.

V The people are as well represented as they deserve to be in parliament. They themselves are the most diversified race of animals in the world, and the House of Commons is the most motley assembly of beings that ever met in the character of legislators. Is there a man in the three kingdoms, beneath the peerage itself, who could not find his counterpart there, and consequently his representative? We want economy, and we want nothing else, to redeem us from all our difficulties; and economy the present ministers are determined to exercise.

W Penny-wise and pound-foolish in truth they are. Let them sacrifice some of their own emoluments, and then they may claim some credit for a disposition to spare the people's pockets.

James Montgomery: From 'A Dialogue of the Alphabet' published in 'Prose by a Poet, Book 1'

The Dutch boors (the farmers on insulated spots of cultivation throughout the colony), were as incapable of comprehending the object of his mission as the barbarians themselves; for it appeared, that at the sacrifice of home, country and friends, all that is dear and desirable in life, this solitary stranger had traversed the land and ocean to fix his abode, where neither wealth was to be accumulated, pleasure pursued, nor honour won; and where, amidst toil, poverty, and contempt, he was about to spend his affections on creatures as insensible as the bushes, and to waste his intellect on minds as barren as the sands.

James Montgomery: From 'An African Valley'

It is a monstrous thing that in this 19th century, within the British dominions a Minister of Christ should be hunted down for teaching the plainest principles of our religion which is declared to be part & parcel of the law of the land.

George Hadfield, M.P. in a letter to James Montgomery on 6th June 1824 concerning the debate in Parliament on the proceedings against Mr. Smith

Christianity and slavery they have long thought, and they now declare it, cannot go together, and in this they are correct enough. Surely this is a theme worthy of the successor of our Cowper!

George Hadfield in a letter to James Montgomery of 5th July 1824

John William Cunningham (1780 - 1861) He was an honorary life-governor of the Church Missionary Society. He wrote *Observations on Friendly Societies.*

George Croly (1780 - 1860) was a poet who wrote *Paris 1815*. He was Rector of St. Stephen's, Walbrook (the London Church from which Chad Varah started the Samaritans twenty-four hour service for the despairing in the 1950s). George Croly however was said to have *a sort of rude and angry eloquence that would have stood him in better stead at the bar than in the pulpit.*

Edward Smedley (1788 - 1836) won the Seatonian Prize for English verse in 1813, 1814, 1827 and 1828. He edited *Encyclopaedia Metropolitana.*

Elliott Cresson (1796 - 1854) was a Quaker merchant and philanthropist living in Philadelphia.

Robert Southey (1774 - 1843) was Poet Laureate. His *The Battle of Blenheim* is a telling indictment against the stupidity of war.

Last week we called the attention of our readers to that *national sin*, Negro-Slavery in the British Colonies, in which, to borrow the energetic language of Mr. Wilberforce in his late admirable appeal on this subject, it is not too much to affirm *that there never was, certainly never before in a Christian country, a mass of such aggravated enormities* . . . in the Britsh territory adjacent to the Cape of Good Hope . . . following advertisement appeared in the Official Gazette . . .

> To be Sold by Auction, to the highest bidder, on Tuesday 15th Oct. next, by order of the Board of Orphan-Masters, on such conditions as will there be specified, the Buildings on the Loan Place Brood Kraal, at Bing River district of Steelenbosch; likewise waggons, ploughs and other instruments of husbandry, cattle, horses, goats and sheep; the whole belonging to the estate of . . . There will also be sold a female slave named Candasa of Mosambique, 54 years old, with her five children, Saphira aged 13 years, Eva, 10, Candasa 9, Jinnotjee, 7, Carolina 5; each to be put up separately, the property of the joint children of H. L. Warnech, deceased.

From Editorial comment in 'The Iris' of 29th April 1823

On 4th June 1823, there was a public meeting of the Auxiliary Missionary Society for the West Riding of Yorkshire held in the Salem Chapel, Leeds. Montgomery wrote *O Spirit of the Living God* for the occasion and, as revised in his *Original Hymns for Christian Worship*, it is still widely sung today to a variety of tunes including NEWBURY (Chetham), GONFALON ROYAL, WINCHESTER NEW (CRASSELIUS (Hamburg); WER NUR DEN LIEBEN GOTT), ST PETERSBURG (RUSSIA (Bortnianski); WELLS (Bortnianski); WELLSPRING), ST BARTHOLOMEW (Duncalf) and MAINZER.

O God! Thou art my God alone;
Early to Thee my soul shall cry,
A pilgrim in a land unknown,
A thirsty land whose springs are dry.

Oh! that it were as it hath been,
When, praying in the holy place,
Thy power and glory I have seen,
And mark'd the footsteps of Thy grace!

Yet through this rough and thorny maze,
I follow hard on Thee, my God!
Thine hand unseen upholds my ways,
I safely tread where Thou hast trod.

Thee, in the watches of the night,
When I remember on my bed,
Thy presence makes the darkness light,
Thy guardian wings are round my head.

Better than life itself Thy love,
Dearer than all beside to me;
For whom have I in heaven above,
Or what on earth, compared with Thee?

Praise with my heart, my mind, my voice,
For all Thy mercy I will give;
My soul shall still in God rejoice,
My tongue shall bless Thee while I live.

James Montgomery: 'Remembrance and Resolution' incorporating minor amendments made in 'Original Hymns for Christian Worship'

O Spirit of the living God!
In all Thy plenitude of grace,
Where'er the foot of man hath trod,
Descend on our apostate race.

Give tongues of fire and hearts of love,
To preach the reconciling word;
Give power and unction from above,
Whene'er the joyful sound is heard.

Be darkness, at Thy coming, light,
Confusion, order in Thy path;
Souls without strength inspire with might.
Bid mercy triumph over wrath.

O Spirit of the Lord! prepare
All the round earth her God to meet;
Breathe Thou abroad like morning air,
Till hearts of stone begin to beat.

Baptize the nations; far and nigh,
The triumphs of the Cross record;
The name of Jesus glorify,
Till every kindred call Him Lord.

God from eternity hath will'd,
All flesh shall His salvation see;
So be the Father's love fulfill'd,
The Saviour's sufferings crown'd through Thee.

James Montgomery: 'The Spirit accompanying the Word of God'

The hymn *Stand up, and bless the Lord* sung to CARLISLE, MONTGOMERY (Goodall), ST MICHAEL (OLD 134th) and DONCASTER (BETHLEHEM (Wesley)) was originally written for the Sheffield Red Hill Wesleyan Sunday School Anniversary on 15th March 1824. The text is based on *Nehemiah* 9, 5 and entitled *The Law read by Ezra and the Covenant renewed.* For the Whitsuntide gathering of the Sheffield Sunday School Union he provided *Sing we the song of those who stand* used today with TIVERTON and NATIVITY (Lahee). The same year he published two volumes under the general title *Prose by a Poet.* Amongst the contents *An African Valley* tells the story of the Hottentot congregation in South Africa. The variety of

Stand up, and bless the Lord,
Ye children of His choice;
Stand up, and bless the Lord your God
With heart, and soul, and voice.

Though high above all praise,
Above all blessings high,
Who would not fear His holy name,
And laud and magnify?

O for the living flame,
From His own altar brought,
To touch our lips, our minds inspire,
And wing to heaven our thought!

There, with benign regard,
Our hymns He deigns to hear;
Though unreveal'd to mortal sense,
The spirit feels Him near.

God is our strength and song,
And His salvation ours;
Then be His love in Christ proclaim'd
With all our ransom'd powers.

Stand up, and bless the Lord,
The Lord your God adore;
Stand up, and bless His glorious name,
Henceforth for evermore.

James Montgomery: 'Exhortation to Praise and Thanksgiving'

Sing we the song of those who stand
Around the eternal throne,
Of every kindred, clime and land,
A multitude unknown.

Life's poor distinctions vanish here;
To-day the young, the old,
Our Saviour and His flock appear
One Shepherd and one fold.

Toil, trial, suffering still await
On earth the pilgrim-throng,
Yet learn we, in our low estate,
The Church-triumphant's song.

Worthy the Lamb for sinners slain,
Cry the redeem'd above,
Blessing and honour to obtain,
And everlasting love.

Worthy the Lamb, on earth we sing,
Who died our souls to save;
Henceforth, O Death, where is thy sting?
Thy victory, O Grave?

Then, Hallelujah! power and praise
To God in Christ be given;
May all who now this anthem raise
Renew the strain in heaven.

James Montgomery: 'The Church Militant learning the Church Triumphant's Song'

subject is evident by these titles: *Old Woman, The Life of a Flower, Juvenile Delinquency, A Dialogue on the Alphabet* and *An Apocryphal Chapter on the History of England.* He also wrote a substantial introduction to an edition of the poems of William Cowper, poet and hymnwriter, who wrote *Hark, my soul! It is the Lord.* He also found time for several poems including *A Sea Piece* written at Bridlington. The members of the Sheffield Philosophical and Literary Society were given the opportunity of subscribing to a painting of Montgomery to be made by a local artist, Thomas Barber.

Food, raiment, dwelling, health and friends,
Thou, Lord, hast made our lot;
With Thee our bliss begins and ends,
As we are Thine, or not.

For these we bend the humble knee,
Our grateful spirits bow;
Yet from thy gifts we turn to Thee:-
Be Thou our portion, Thou!

James Montgomery: 'The Family Altar' as slightly revised for his definitive version in 'Original Hymns for Christian Worship'

Be known to us in breaking bread,
But do not then depart;
Saviour, abide with us, and spread
Thy table in our heart.

There sup with us in love divine;
Thy body and Thy blood,
That living bread, that heavenly wine,
Be our immortal food.

James Montgomery: 'The Family Table'

. . . *Table Talk*, is a dialogue in verse, of which the subjects are chiefly the commonplace politics of the day; and the Author, by an easy pedestrian pace, has got midway through his theme before he kindles into any thing like fury, or betrays any strong symptom of the diviner mood. Then, indeed, comes a glorious burst, in which the patriot, the Christian, and the bard all unite in a warning, sufficient to alarm the most supine statesman, touching the real perils, and false security of a nation hastening unconsciously to ruin, through the undermining vices of luxury and licentiousness.

'They trust to navies, and their navies fail, –
God's curse can cast away ten thousand sail!
They trust in armies, and their courage dies;
In wisdom, wealth, in fortune, and in lies:
But all they trust in withers, as it must,
When He commands, in whom they place no trust.
Vengeance at last pours down upon their coast
A long despised, but now victorious, host;
Tyranny sends the chains, that must abridge
The noble sweep of all their privilege;
Gives liberty the last, the mortal shock;
Slips the slave's collar on, and snaps the lock'

From an introductory essay by James Montgomery to an edition of the 'Poems of William Cowper Esq'

The discontinuance of religious ordinances necessarily broke up Sunday-schools, and dissolved all National, Lancasterian, and charitable institutions, in which the children of the poor are taught to become good servants, good neighbours and good subjects. Hereupon some hundreds of thousands of naughty boys and girls were turned loose upon one another and upon the community; the plague of which was more particularly felt on the Sabbath days, – when neither the frogs, the flies, nor the locusts of Egypt were more pestilent in their day and generation, than were these young fry.

James Montgomery: From 'An Apocryphal Chapter in the History of England' published in 'Prose by a Poët, Book 2'

The Christian Psalmist, published in 1825, contained 562 hymns including 101 by Montgomery. Some other authors he chose for inclusion were Charles Wesley, Augustus Toplady, John Newton, Benjamin Beddome, Philip Doddridge, Anna Barbauld, Reginald Heber, Simon Browne, William Cowper and Jane Taylor. The volume included the first publication of *Be known to us in breaking bread* under the title *The Family Table* which is sung to BELMONT, TALLIS' ORDINAL (TALLIS NINTH TUNE; TALLIS) and BALLERMA. It was immediately preceded by a text

I am really so low in spirit as well as feeble in body, that I cannot seize up courage to accept the obliging invitation of your literary and philosophical friends to attend their anniversary . . .

James Montgomery writing to Dr. James Williamson in Leeds, 21st May 1825

. . . Olney Hymns ought to be for ever dear to the Christian Public, as an unprecedented memorial, in respect to its Authors, of the prove of divine grace, which called one of them from the negro slave-market, on the coast of Africa, to be a burning and a shining light to the Church of God at home, – and raised the head of the other, when he was a companion of lunatics, to make him, (by a most mysterious dispensation of gifts,) – a poet of the highest intellectuality, and in his song an unshaken, uncompromising confessor of the purest doctrines of the Gospel, even when he himself had lost sight of its consolations.

From James Montgomery's Preface to an edition of the 'Olney Hymns'

The morn was beautiful, the storm gone by;
Three days had pass'd; I saw the peaceful main,
One molten mirror, one illumined plane,
Clear as the blue, sublime, o'erarching sky:
On shore that lonely vessel caught mine eye,
Her bow was seaward, all equipt her train,
Yet to the sun she spread her wings in vain,
Like a caged Eagle, impotent to fly;
There fix'd as if for ever to abide;
Far down the beach had roll'd the low neap-tide:
Whose mingling murmur faintly lull'd the ear:
'Is this', methought 'is this the doom of pride,
Check'd in the onset of thy brave career,
Ingloriously to rot by piecemeal here?'

James Montgomery: 'Sonnet II' from 'A Sea Piece'

Dr. Watts may almost be called the inventor of hymns in our language; for he so far departed from all precedent, that few of his compositions resemble those of his fore-runners, – while he so far established a precedent to all his successors, that none have departed from it, otherwise than according to the peculiar turn of mind in the writer, and the style of expressing Christian truths employed by the denomination to which he belonged. Dr. Watts himself, though a conscientious dissenter, is so entirely catholic in his hymns, that it cannot be discovered from any of these, (so far as we can recollect), that he belonged to any particular sect; hence, happily for his fame, or rather, it ought to be said, happily for the Church of Christ, portions of his psalms and hymns have been adopted in most places of worship where congregational singing prevails . . .

It might be expected, however, that in the first models of a new species of poetry, there would be many flaws and imperfections, which later practitioners would discern and avoid. Such, indeed, are too abundant in Dr. Watt's Psalms and Hymns; and the worst of all is, that his authority stands so high with many of his imitators, that, while his faults and defects are most faithfully adopted, his merits are unapproachable by them. The faults are principally prosaic phraseology, rhymes worse than none, and none where good ones are absolutely wanted to raise the verse upon its feet, and make it go, according to the saying *on all-fours*; though, to do the Doctor justice, the metre is generally free and natural, when his lines want every other qualification of poetry.

From James Montgomery's Introduction to 'The Christian Psalmist'

Isaac Watts

ntitled *The Family Altar*. Montgomery's *Our soul shall ɲagnify the Lord*, sung now to OLD 100TH, was designated *'or a Female Friendly Society*. He was consulted in 1831 by ᐟIr. H. B. Tymby, the Mayor of Worcester, about setting up here what the Irish call *An Old Women's Society*. Another rom the collection still sung to-day is *According to thy racious word*. Tunes are BANGOR, ABBEY, and TALLIS')RDINAL (TALLIS NINTH TUNE; TALLIS). He had for ome time become weary of running a newspaper and ɔrinting business and sold *The Iris* to John Blackwell who ıad resigned from the Methodist Ministry on account of ll health. The price was around £1,300. John Holland was ɩppointed the Editor. He went on to edit the *Newcastle Courier*, returning in 1837 to edit the *Sheffield Mercury* vhich was absorbed in 1848 by the *Sheffield Independent*. The *Sheffield Iris* as it was renamed itself continued to be ɔublished until 1848. When Montgomery first arrived in Sheffield, the establishment despatched him to prison. On his 'etirement from business, one hundred and sixteen gentlemen ıonoured him, with an inevitable apology for an earlier nisjudgement, with a public dinner at the Tontine Inn. The ɔccasion was his birthday on 14th November and the event in ɩpprobation of his public and private virtues was at the Tontine Inn with Lord Milton, M.P. in the chair and the ɔther guests seated near the seat of honour were Reverend Dr. Milner of Thriberg, Hugh Parker Esq., the respected nagistrate and Robert Chaloner, the Member of Parliament for the City of York. In all there were about one hundred guests. The same day, the Reverend S. Langton preached in connection with the Montgomery Friendly Society at the Parish Church.

B. Huntsman & Co, Sheffield - Steel Furnaces & Offices, c. 1830

Our purpose here is to do honour to the individual, who, in his whole course of life, ever made it an object to promote peace.

— — — — —

Many years ago, it might have been objected against our friend, that he had been rendered answerable to the laws for certain alleged offences; but these imputed offences are forgotten, and he has proved by his subsequent life that the impugners of his principles were mistaken.

— — — — —

I have had recent proofs of the lively interest which he takes in the Great School Establishments in this town.

— — — — —

If we do not know the value of religion, we cannot by any means administer to the benefit and comfort of mankind.

— — — — —

The town of Sheffield has been exalted by his literary attainments.

Extracts from the speech by the Chairman, Lord Milton, M.P. on the occasion of the Public Dinner given to James Montgomery

Our souls shall magnify the Lord,
In Him our spirit shall rejoice;
Assembled here with sweet accord,
Our hearts shall praise Him with our voice.

Since He regards our low estate,
And hears His handmaids when they pray,
We humbly plead at Mercy's gate,
Where none are ever turn'd away.

The poor are His peculiar care,
To them His promises are sure;
His gifts the poor in spirit share:
O may we always thus be poor!

God of our hope, to Thee we bow,
Thou art our refuge in distress;
The Husband of the widow, Thou,
The Father of the fatherless.

May we the law of love fulfill,
To bear each other's burdens here;
Suffer, and do Thy righteous will,
And walk in all Thy faith and fear.

Didst Thou not give Thy Son to die,
For our transgressions, in our stead?
And can Thy goodness ought deny
To those for whom Thy Son hath bled?

Then may our union, here begun,
Endure for ever, firm and free,
At Thy right-hand may we be one,
One with each other, and with Thee.

James Montgomery: 'For a Female Friendly Society'

According to thy gracious word,
In meek humility,
This will I do, my dying Lord:
I *will* remember thee.

Thy body, broken for my sake,
My bread from heaven shall be;
Thy testamental cup I take,
And thus remember Thee.

Gethsemane can I forget?
Or there Thy conflict see,
Thine agony and bloody sweat,
And not remember Thee?

When to the cross I turn mine eyes
And rest on Calvary,
O Lamb of God, my sacrifice!
I must remember Thee: –

Remember Thee, and all Thy pains,
And all Thy love to me;
Yea, while a breath, a pulse remains,
Will I remember Thee.

And when these failing lips grow dumb,
And mind and memory flee,
When Thou shalt in Thy Kingdom come,
Jesus, remember me.

James Montgomery as printed in
'Original Hymns for Christian Worship'

I recollect, that I once went into Derbyshire, in company with a friend, and a niece of mine, – a young person, born and brought up on the banks of the Thames; accustomed to the gay, populous, cultivated scenery of Kent and Middlesex, who scarcely ever had seen a common less frequented than Blackheath, or an eminence more rugged than Shooter's-hill. She was sufficiently lively and fluent of tongue till we had got upon the High Moors; then she grew gradually serious, and at length silent even to sadness. The magnificence of nature in a new form overawed her; the loneliness of the moorlands made thought retire inward; the weight of the mountains seemed to lie upon her spirit, and the depth of the valleys, as we approached Hathersage, to absorb all consciousness, except that of their own dreadful, but delightful presence. Wonder, admiration, and transport, were sublimed by terror. Sometime afterwards, talking of that morning's excursion, I said to her 'Betsey, what did you think of the Peak mountains, when we were among them?' – 'Oh!' she replied, with great simplicity, 'I wanted to be quite still; I wished that nobody would speak to me.' – A measure of this deep, undefinable feeling has possessed me in the anticipation of this day, and amidst the festivity of this scene; I could have wished, had it been possible, that I might have been silent, and even invisible among you . . .

– – – – –

No; for, to the limits of temperance, where, I suppose, enjoyment ends, I, too, can enjoy the luxury and exhilaration of a well spread board, surrounded with good company, when I have nothing else to do *but* to enjoy them.

Extracts from a speech by James Montgomery
at a farewell dinner given in his honour

And this was my ground, – a plain determination, come wind or sun, come fire or flood, to do what was right. I lay stress on the purpose, not the performance, for this was the polar star to which my compass pointed, though with considerable *variation of the needle*. Through characteristic weakness, perversity of understanding, or self-sufficiency, I have often erred, failed, and been overcome by temptation, on the wearisome pilgrimage through which I have toiled, – now struggling through the Slough of Despond, then fighting with evil spirits in the Valley of Humiliation, more than once escaping martyrdom from Vanity Fair, and once, at least, (I will not say when,) a prisoner in Doubting Castle, under the discipline of Giant Despair. Now, though I am not writing this Address in one of the shepherds' tents on the Delectable Mountains, yet, like Christian, from that situation, I can look back on the past, with all its anxieties,

Phrenology, as a science, must stand or fall by facts, of which, if there be yet too few to decide its legitimacy, there are far too many, of plausible bearing, to allow it to be laughed out of credit, except by the prejudiced and superficial, with whom it would be no credit to be otherwise treated. The object of the following Essay is to show, that were it established to the utmost claims of its reasonable advocates, it involves no fatality in its issues, because all the primal dispositions which it indicates by their respective signs, may be converted to *useful*, or perverted to *evil* purposes; and that not in individuals only, but in whole nations. nor for a brief period only, but through a succession of ages.

James Montgomery: 'On the Phrenology of the Hindoos and Negroes'

Had the Essayist been acquainted with the natural laws of organized beings, – had he understood any thing of physiology, or even the common fundamental principles of the very science which he labours to overthrow, – he would certainly have been less dogmatical on the occasion. The Hindoos are mentally imbecile. Now, according to the doctrines of phrenology, this may depend on a lymphatic constitution, a defective cerebral organization, or on both causes operating at once. The skulls of these people, which are met with in European collections, are remarkable for their smallness; and we fearlessly maintain, that if such be common to the nation at large, the majority of individuals will not fail to betray great weakness of mind.

Gordon Thompson: 'Strictures on Mr. Montgomery's Essay on the Phrenology of the Hindoos and Negroes'

trials, and conflicts, – thankful that it *is* the past. Of the future I have little foresight, and I desire none with respect to this life, being content that *shadows, clouds, and darkness rest upon it*, if I yet may hope, that *at even-tide there will be light.*

From Mr. Montgomery's farewell Address, as published in his last 'Iris'

Sarah & Elizabeth Gales & James Montgomery

Montgomery's retirement did not mean the end of his association with Sheffield. All these years he had continued to live in the home he shared with the Misses Gales; and he continued to do so until in 1836, when on account of one of the sister's ill health, he moved to The Mount, still in Sheffield. In 1826, he again visited Scarborough, a favourite sea-side resort of his.

On 7th February 1827, Montgomery read a paper before the Literary and Philosophical Society of Sheffield, of which he was an active member. The subject was *The Phrenology of the Hindoos and Negroes; showing* that the actual *Character of nations as well as of individuals, may be modified by moral, political and other circumstances, in direct contradiction to their cerebral developments*. The paper was later printed and brought an angry rebuff from a Lecturer on Physiology, and on the Nature and Treatment of Diseases, at the Sheffield School of Anatomy and Medicine. On Easter Monday 1827, the stone-laying ceremony took place for Sheffield's Music Hall.

The same year, Montgomery wrote the narrative poem *The Pelican Island* suggested by a passage in Captain Flinder's *Voyage to Terra Australis*. He describes his poem as *Truth severe by fairy-fiction drest*. August 13th was the centenary of the memorable day in 1727 when there was a union of all hearts and minds in the congregation of the United Brethren at Herrnhüt amongst whom, since their revival five years previously, considerable differences had prevailed on minor points. Their reconciliation at the Holy Sacrament in the Parish Church of Bethelsdorf was the inspiration a century later for Montgomery's *The God of your forefathers praise*. His portrait was painted again this time by John Jackson.

Wading through marshes, where the rank sea-weed
With spongy moss and flaccid lichens strove,
Flamingoes, in their crimson tunics, stalk'd
On stately legs, with far-explorng eye;
Or fed and slept, in regimental lines,
Watch'd by their sentinels, whose clarion-screams
All in an instant woke the startled troop,
That mounted like a glorious exhalation,
And vanish'd through the welkin far away,
Nor paused till, on some lonely coast alighting,
Again their gorgeous cohort took the field.

The fierce sea-eagle, humble in attire,
In port terrific, from his lonely eyrie
(Itself a burden for the tallest tree)
Look'd down o'er land and sea as his dominions:
Now, from long chase, descending with his prey,
Young seal or dolphin, in his deadly clutch,
He fed his eaglets in the noonday sun:
Nor less at midnight ranged the deep for game;
At length entrapp'd with his own talons, struck
Too deep to be withdrawn, where a strong shark,
Roused by the anguish, with impetuous plunge,
Dragg'd his assailant down into the abyss,
Struggling in vain for liberty and life;
His young ones heard their parent's dying shrieks,
And watch'd in vain for his returning wing.

– – – – –

Millions of creatures such as these, and kinds
Unnamed by man, possess'd those busy isles;
Each, in its brief existence, to itself,
The first, last being in the universe,
With whom the whole began, endured, and ended:
Blest ignorance of bliss, not made for them!
Happy exemption from the fear of death,
And that which makes the pangs of death immortal,
The undying worm, the fire unquenchable,
– Conscience, the bosom-hell of guilty man!
The eyes of all look'd up to Him, whose hand
Had made them, and supplied their daily need;
Although they knew Him not, they look'd to Him;
And He, whose mercy is o'er all his works,
Forgot not one of his large family,
But cared for each as for an only child.

James Montgomery: From 'The Pelican Island'

Poetry is certainly something more than good sense, but it must be good sense at all events; just as a palace is more than a house, but it must be a house, at least.

Samuel Taylor Coleridge, with whom Montgomery corresponded, writing in 'Table Talk', 9th May 1830

They walk'd with God in peace and love,
But fail'd with one another;
While sternly for the faith they strove,
Brother fell out with brother:
But He, in whom they put their trust,
Who knew their frames, that they were dust,
Pitied and heal'd their weakness.

He found them in His house of prayer,
With one accord assembled,
And so reveal'd His presence there.
They wept for joy, and trembled;
One cup they drank, one bread they broke,
One baptism shared, one language spoke,
Forgiving and forgiven.

James Montgomery: Verses from 'The God of your forefathers praise'

An interesting assignment about this time was to write an introductory essay to Benjamin Wisner's *Memoirs of the late Mrs S. Huntingdon of Boston, Mass.* whose diaries and correspondence were published to show the value of the actions of an ordinary Christian lady.

During 1828 he toured Wales, visiting Chepstow, Milford Haven and Aberystwyth. He climbed the Brecon Beacons but all he saw when attempting Snowdon was thick mist and cloud. The poet Felicia Hemans, on whom he called described him as *very pleasing in manner and countenance, notwithstanding a mass of troubled, streaming, metreonic-looking hair, that seemed as though it had just been contending with the blasts of Snowdon, from which he had just returned, full of animation and enthusiasm.*

Now, poetry is a school of sculpture, in which the art flourishes, not in marble or brass, but in that which outlasts both, – in letters, which the fingers of a child may write or blot, but which, once written, Time himself may not be able to obliterate; and in sounds, which are but passing breath, yet being once uttered, by possibility, may never cease to be repeated. Sculpture to the eye, in palpable materials, is of necessity confined to a few forms, aspects, and attitudes. The poet's images are living, breathing, moving creatures; they stand, walk, run, fly, speak, love, fight, fall, labour, suffer, die, – in a word, they are men of like passions with ourselves, undergoing all the changes of actual existence, and presenting to the mind of the reader, solitary figures, or complicated groups, more easily retained (for words are better recollected than shapes substances), and infinitely more diversified, than the chisel could hew out of all the rocks under the sun.

James Montgomery: From 'The Pre-Eminence of Poetry', a lecture delivered at the Royal Institution

At the beginning of 1829, in a preface to a new edition of *Olney Hymns* he wrote in reference to William Cowper and John Newton, *There are few joint-memorials of friendship and talents, raised by kindred spirits in polite literature.* For that year's Sheffield Sunday School Union Anniversary in June he wrote *Palms of glory, raiment bright*, still occasionally sung to PALMS OF GLORY. The same year, the poet Thomas Campbell sought a contribution to an anthology to raise funds for the relief of Italian and Spanish refugees, and Thomas Pringle, on behalf of the Anti-Slavery Society, commissioned a poem about South Africa, from where he was virtually banned after his newspapers had been closed down. The following year, Montgomery delivered a series of Lectures on Poetry and General Literature at the Royal Institution but he did not have the same flair for public speaking as he did for writing. Part of the fees for the lectures remained unpaid and became the subject of considerable correspondence. He also designed the seal for the Sheffield Philosophical and Literary Society. Some credit the honour of starting the Sunday School movement to a Methodist, Hannah Ball: but Robert Raikes, who has earned the title more widely, if less accurately, led to Montgomery's text, *Let songs of praise arise* for Robert Raikes' birthday on September 4th 1831, for the Sunday School Jubilee. He also compiled *Voyages and Travels round the World by the Reverend Daniel Tyerman and George Bennett Esq., deputed from the London Missionary Society to visit their various stations in the South Sea Islands, Australia, China, India, ladagascar and South Africa between the years 1821 and '829.* James always used every opportunity to remind isteners he was the son of missionaries and he must have ound these journals fascinating reading. A poem written that ear takes a distinctly non-sentimental view in *A Benediction or a Baby*.

What, in the labour, pain, and strife,
Combats and cares of daily life?
– In *his* cross-bearing steps to tread,
Who had not where to lay his head.
James Montgomery: From 'A Benediction for a Baby'

We ought to cultivate a cheerful view of all the providences of God . . . It is a great attainment in Christian wisdom to be able to discover those lucid spots in the cloudy atmosphere which envelopes us, in the present state; to give God the praise for them, and take to ourselves, and impart to others, the comfort of them.
From the writings of Mrs. S. Huntingdon, 1820

In Mrs. Huntingdon we have an exemplification of Christian character in the female sex, rising into grace, expanding into beauty, and flourishing in usefulness, from infancy to youth, and from youth to womanhood; *then*, without reaching old age, translated to Paradise, *like a tree planted by the rivers of water*, that brought forth its fruit in due season, and whose leaf also withered not; being cut down in its prime, and remembered only as the glory of the place where it grew. There were no extraordinary incidents in her brief existence; she occupied no eminent station in society; she was endowed with no splendid talents; but on account of these very deficiencies, (defects they were not) something *more excellent*, yet *attainable by all*, having been found in her, she may be presented as a model to others passing through the same ordinary circumstances, whereby they may form themselves to meet every change, till the last; and in that last, be perfectly prepared for a state beyond the possibility of change for ever.
From an introductory essay by James Montgomery to Benjamin Wisner: 'Memoirs of the late Mrs. S. Huntingdon of Boston, Mass.'

Resolved, That the cordial thanks of this Committee be presented to James Montgomery Esq., for the Services he has rendered during the current year in visiting Auxiliary and Branch Societies.
Minute of the Committee of the British and Foreign Bible Society, 1st November 1830

His hymn *Holy, holy, holy Lord*, which he eventually placed first in his definitive collection *Original Hymns for Christian Worship*, his text linked to *Isaiah* 6, 3, was written in 1832 and published in the *Congregational Hymn Book*, 1836. Tunes used include SALZBURG (Hintze) (ALLE MENSCHEN MÜSSEN STERBEN; SCHÖNBERG) and TICHFIELD (GOD OF GLORY; TITCHFIELD). In February 1832, he was travelling with a blind friend from Gloucester to Tewkesbury, when he noticed a field *which had been ploughed recently but not apparently harrowed, for the surface lay not in furrows . . . But upon it were several women and girls in rows, one behind another, literally, as though they were engaged in parallel lines, but did not keep*

. . . Or if you would agree to meet me any day in the City & go home with me to a family dinner – (a small pot-luck entertainment, such as a poor poet may offer to his venerated sire of the craft), – you would still more especially oblige me. I will ask no other party to meet you, both because I am not rich enough to give parties, and also because I wish to enjoy your conversation quietly at my family fireside.
Thomas Pringle in a letter to James Montgomery on 25th May 1831

Holy, Holy, Holy Lord,
God of Hosts! when heaven and earth,
Out of darkness at Thy word,
Issued into glorious birth,
All Thy works before Thee stood,
And Thine eye beheld them good,
While they sang with sweet accord,
Holy, Holy, Holy Lord!

James Montgomery: The opening verse of 'Thrice Holy' as printed in 'Original Hymns for Christian Worship'

Praise the Lord through every nation;
His holy arm hath wrought salvation;
Exalt Him on His Father's throne:
Praise your king, ye Christian legions,
Who now prepares, in heavenly regions,
Unfailing mansions for His own:
With voice and minstrelsy,
Extol His majesty;
Hallelujah!
His praise shall sound – all nature round,
Where'er the race of man are found.

James Montgomery: From 'For Ascension Day', the opening verse of a hymn paraphrased from the Dutch in its original metre. Sung to WACHET AUF! (SLEEPER AWAKE)

Sow in the morn thy seed,
At eve hold not thine hand;
To doubt and fear give thou no heed,
Broad-cast it o'er the land.

Beside all waters sow,
The highway furrows stock,
Drop it where thorns and thistles grow,
Scatter it on the rock.

The good, the fruitful ground,
Expect not here nor there;
O'er hill and dale, by plots, 't is found;
Go forth, then, every where.

Thou know'st not which may thrive
The late or early sown;
Grace keeps the precious germs alive
When and wherever strown.

And duly shall appear,
In verdure, beauty, strength;
The tender blade, the stalk, the ear,
And the full corn at length.

Thou canst not toil in vain;
Cold, heat, and moist, and dry,
Shall foster and mature the grain,
For garners in the sky.

Thence, when the glorious end,
The day of God is come,
The angel-reapers shall descend,
And heaven cry, – *Harvest Home!*

James Montgomery: 'The Field of the World' from 'A Poet's Portfolio'

ace with each other in their work. He vondered what they were about but his lind companion conjectured – they vere dibbling. He explained this was a ecessary procedure in the harsh conomic conditions because only a hird of the seed was needed by omparison with ordinary means of owing: holes are picked in lines along he field, and each hole received just ts proper ration. Montgomery, for his art, considered dibbling to be very npoetical and unpicturesque – there is either grace of motion or attitude to it. Give him *broadcast* sowing every time, cattering the seeds on the right hand nd on the left, in liberal handfuls. By he time the journey was completed, *ow in the morn thy seed* was written. It as been sung to ALDERSGATE and SWABIA. The same year he wrote *Sing Hallelujah; sing* as a ymn of thanksgiving for the removal of the cholera from Sheffield. One of Montgomery's correspondents was Thomas Raffles, who preached a sermon under the title *The mingled Character of the Divine Dispensations recognised and cknowledged* at the Scotch Session Church, Mount Pleasant, Liverpool on 1st January 1833, in which he declared with ertainty that the cholera had been a visitation from God for he people's sins.

The Lord put forth His hand,
He touch'd us and we died;
Vengeance went through the land,
But mercy walk'd beside;
He heard our prayers; He saw our tears,
And stay'd the plague, and quell'd our fears.

James Montgomery: From 'Sing Hallelujah; sing'

Palms of glory, raiment bright,
Crowns that never fade away,
Gird and deck the saints in light,
Priests, and kings, and conquerors they.

James Montgomery: The first verse of 'Heaven in prospect'

Though poor and mean the place,
And small the band he taught;
Millions since then have shared the grace;
Behold what God hath wrought.

James Montgomery: From 'Let songs of praise arise'

On 23rd January 1833 he wrote *Pour out Thy Spirit from on igh* for the Reverend J. Birchall, the Rector of Newbury, who was publishing a selection of hymns. It was repeated in Edward Bickersteth's *Christian Psalmody* of the same year. It is sometimes sung with the first line altered to *Lord, pour hy Spirit from on high.* Tunes include DUKE STREET, WAINWRIGHT, SONG 34, WARRINGTON, MAINZER, MELCOMBE and LUDBOROUGH. In 1834, the poet wrote *In the hour of trial* and John Ellerton, the author of *The day thou gavest Lord is ended,* told its story in his notes to *Church Hymns.* A tune is BOHEMIA. Amongst the 1834 crop of poetry is *The Sunflower.* As President of Sheffield Mechanics' Institute, Montgomery secured the services of the Lord Chancellor, Lord Henry Brougham to address them. He was himself that year's invited speaker at the Cutlers' Feast.

That this visitation was from God, in a most marked and especial manner, I shall not stay to prove. With the sceptic and the infidel, who would of course dispute or deny it, I will have in this discourse no argument. I take it for granted that all in this assembly are ready to admit a fact, so obvious that the most thoughtless and superficial were constrained to perceive and to acknowledge it – that the pestilence was directly and especially from God!

Thomas Raffles: From 'The Mingled Character of the Divine Dispensations recognised and acknowledged'

In 1835, Montgomery published *A Poet's Portfolio.* Its first book comprised a group of narrative poems, including *Lord*

Within Thy temple, when we stand,
To teach the truth, as taught by Thee,
Saviour, like stars in Thy right-hand,
The angels of the Churches be.

James Montgomery: Verse 2 of 'For a Meeting of Ministers' which begins 'Pour out Thy Spirit from on high'

In the hour of trial,
Jesus, pray for me,
Lest, by base denial,
I depart from Thee:
When Thou seest me waver,
With a look recall,
Nor, for fear or favour,
Suffer me to fall.

James Montgomery: Verse 1 of 'Prayers on Pilgrimage'

Falkland's Dream. Lord Falkland wa Secretary of State to King Charles anc Montgomery prints as a preface a substantial piece of Clarendon's *Histor of the Rebellion. The Patriot's Pass-Wor* was concerned with *the achievemen of Arnold de Winkelried, at the battle o Sempach, in which the Swiss insurgent secured the freedom of their country against the power of Austria, in th fourteenth century. The Voyage of th Blind* tells the story of the *Leon* a Montgomery imagines it. It is based o an account of *Le Rodeur*, many o whose crew and cargo of Africa negroes, all bound for the West Indies were struck by blindness, aggravated b dysentry. In their own trouble, the crew of *Le Rodeur* refused help to the *Leon*, whose whol complement had lost their sight. *An Every-Day Tale* wa written for a benevolent society in the metropolis, th object of which is to relieve poor women during the firs month of their widowhood, to preserve what little propert they may have from wreck and ruin, in a season o embarrassment, when kindness and good counsel ar especially needed; and, so far as may be practicable, t assist the destitute with future means of maintainin themselves and their fatherless children. His *A Tale Withou a Name* was sparked off by words in Scott's *Marmion*, whils *Ugolino and Ruggieri* owes its creation to Dante's *Inferno*.

The second line in the first verse has given rise to much controversy, as to whether a request for the prayers of our Lord now is a legitimate inference from St. Luke xxii.32. Those who think otherwise would appear to press St. John xvi.26 beyond its fair limits. Compare Hebrews ii. 17, 18, with iv. 14-16.

John Ellerton, in the Annotated Edition of 'Church Hymns', comments on James Montgomery's 'In the hour of trial'

Among the miscellaneous poems in the second book i *Birds* in which the author has a series of conversations wit the swallow, skylarks, the cuckoo, the red-breast, th sparrow, the ring-dove, the nightingale, the water-wagtail the wren, the thrush, the blackbird, the bullfinch, th goldfinch, the grey linnet, the red linnet, the chaffinch the canary, the tomtit, the swift, the king-fisher, th woodlark, the cock, the jack-daw, the bat, the owl, rooks the jay, the peacock, the swan, the pheasant, the raven the parrot, the magpie, the corn-crake, the stork, th woodpecker, the hawk, vultures, the humming bird, th eagle, the pelican, the heron, the bird of paradise and th ostrich.

**Eagle of flowers! I see thee stand,
And on the sun's noon-glory gaze . . .**

James Montgomery: From 'The Sun-Flower'

A Cry from South Africa is concerned with the building o a chapel for the Negro Slaves in Cape Town, in 1828. *The Cholera Mount* was written during the 1832 epidemic whe victims were required to be buried in specially appointe laces. *Songs on the Abolition of Negro Slavery in the British olonies* on 1st August 1834, provide a group of five poems. he third book of *A Poet's Portfolio* brings together various eligious verses, some suitable as hymns, and includes a longer iece *The Chronicle of Angels. Glad was my heart to hear,* ased on Psalm 122, is sung to FALCON STREET (SILVER TREET) and RHODES.

War, civil war, was raging like a flood,
England lay weltering in her children's blood;
Brother with brother waged unnatural strife,
Sever'd were all the charities of life:
Two passions, – virtues they assumed to be, –
Virtues they *were*, – romantic loyalty,
And stern, unyielding patriotism, possess'd
Divided empire in the nation's breast;
As though two hearts might in one body reign,
And urge conflicting streams from vein to vein.

James Montgomery: From 'Lord Falkland's Dream' printed in 'A Poet's Portfolio'

When there was any overture or hope of peace he would be more erect and vigorous, and exceedingly solicitous to press any thing which he thought might promote it; and, sitting among his friends, often, after a deep silence, and frequent sighs, would, with a shrill and sad accent, ingeminate the word *Peace! peace!* and would profess that the very agony of the war, and the view of the calamities and desolation the kingdom did and must endure, took his sleep from him, and would shortly break his heart.

From the extract about Lord Falkland in 'Edward Hyde Clarendon: History of the Rebellion' with which Montgomery prefaced his poem 'Lord Falkland's Dream'

Shuddering humanity asks, *Who are these?*
And what their crime? – The fell by one disease!
By the blue pest, whose gripe no art can shun,
No force unwrench, out-singled one by one;
When, like a monstrous birth, the womb of fate
Bore a new death of unrecorded date,
And doubtful name – Far east the fiend begun
Its course; thence round the world pursued the sun,
The ghosts of millions following at its back,
Whose desecrated graves betray'd their track.
On Albion's shores unseen the invader stept,
Secret and swift through field and city swept;
At noon, at midnight, seized the weak, the strong,
Asleep, awake, alone, amid the throng;
Kill'd like a murderer; fix'd its icy hold,
And wrung out life with agony of cold;
Nor stay'd its vengeance where it chrush'd the prey,
But set a mark, like Cain's, upon their clay,
And this tremendous seal impress'd on all,
Bury me out of sight and out of call.

James Montgomery: From 'The Cholera Mount: Lines on the Burying-Place for Patients who died of Cholera Morbus; a Pleasant Eminence in Sheffield Park'

. . . So still, so dense, the Austrian stood,
A living wall, a human wood.
Impregnable their front appears,
All-horrent with projected spears,
Whose polish'd points before them shine,
From flank to flank, one brilliant line,
Bright as the breakers' splendours run
Along the billows to the sun.

James Montgomery: From 'The Patriot's Pass-Word'
printed in 'A Poet's Portfolio'

. . . Scarce was he buried out of sight,
Ere his tenth infant sprang to light,
And Mary, from her child-bed throes,
To instant, utter ruin rose;
Harvests had fail'd, and sickness drain'd
Her frugal stock-purse, long retain'd;
Rents, debts, and taxes all fell due,
Claimants were loud, resources few,
Small, and remote; – yet time and care
Her shatter'd fortunes might repair,
If but a friend, – a friend in need, –
Such friend would be a friend indeed, –
Would, by a mite of succour lent,
Wrongs irretrievable prevent!
She look'd around for such an one,
And sigh'd, but spake not, – *Is there none?*
– Oh! if he come not ere an hour,
All will elapse beyond her power,
And homeless, helpless, hopeless, lost,
Mary on this cold world be tost
With all her babies! * * * * *

Came such a friend? – I must not say;
Mine is a tale of every day:
But wouldst thou know the worst of all,
The wormwood mingled with the gall,
Go visit thou, in their distress,
The widow and the fatherless,
And thou shalt find such woe as this,
Such breaking up of earthly bliss,
Is no strange thing, – but, strange to say!
The tale – the truth – of every day.

Go visit *thou*, in their distress,
The Widow and the Fatherless.

James Montgomery: From 'An Every-Day Tale'
printed in 'A Poet's Portfolio'

Wren, canst thou squeeze into a hole so small?
– Ay, with nine nestlings too, and room for all;
Go, compass sea and land in search of bliss,
Then tell me if you find a happier home than this.

James Montgomery: 'The Wren' from 'Birds' printed in 'A Poet's Portfolio'

Dost thou not languish for thy father-land,
Madeira's fragrant woods and billowing strand?
– My cage is father-land enough for me;
Your parlour all the world, – heaven, earth, and sea.

James Montgomery: 'The Canary' from 'Birds' printed in 'A Poet's Portfolio'

Sparrow, the gun is levell'd, quit that wall.
– Without the will of heaven I cannot fall.

James Montgomery: 'The Sparrow' from 'Birds' printed in 'A Poet's Portfolio'

Stock-still upon that stone, from day to day,
I see thee watch the river for thy prey.
– Yes, I'm the tyrant here: but when I rise,
The well train'd falcon braves me in the skies;
Then comes the tug of war, of strength and skill,
He dies, impaled on my updarted bill,
Or, powerless in his grasp, my doom I meet,
Dropt as a trophy at his master's feet.

James Montgomery: 'The Heron' from 'Birds' printed in 'A Poet's Portfolio'

Good angels still conduct, from age to age,
Salvation's heirs, on nature's pilgrimage;
Cherubic swords, no longer signs of strife,
Now point the way, and keep the tree of life;
Seraphic hands, with coals of living fire,
The lips of God's true messengers inspire;
Angels, who see their heavenly Father's face,
Watch o'er his little ones with special grace;
Still o'er repenting sinners they rejoice,
And blend their myriad-voices as one voice.
Angels, with healing virtue in their wings,
Trouble dead pools, unsluice earth's bosom-springs,
Till fresh as new-born life the waters roll;
Lepers and lame step in and are made whole.

James Montgomery: From 'The Chronicle of Angels' printed in 'A Poet's Portfolio'

Curst was her trade and contraband,
Therefore that keel, by guilty stealth,
Fled with the darkness from the strand,
Laden with living bales of wealth:
Fair to the eye her streamers play'd
With undulating light and shade;
White from her prow the gurgling foam
Flew backward tow'ds the negro's home,
Like his unheeded sighs;
Sooner that melting foam shall reach
His inland home, than yonder beach
Again salute his eyes.

. . . . .

There came an angel of eclipse,
Who haunts at time the Atlantic flood,
And smites with blindness, on their ships,
The captives and the men of blood.
– *Here*, in the hold the blight began,
From eye to eye contagion ran;
Sight, as with burning brands, was quench'd;
None from the fiery trial blench'd,
But, panting for release,
They call'd on death, who, close behind,
Brought pestilence to lead the blind,
From agony to peace.

James Montgomery: Extracts from 'The Voyage of the Blind' printed in 'A Poet's Portfolio'

Glad was my heart to hear
My old companions say,
Come, – in the House of God appear,
For 'tis an holy day.

Thither the tribes repair,
Where all are wont to meet,
And joyful in the House of Prayer
Bend at the Mercy-seat.

For friends and brethren dear,
Our prayer shall never cease,
Oft as they meet for worship here,
God send his people peace.

James Montgomery: Verses 1, 3 and 6 of 'For the Peace and Prosperity of the Church'

In 1835, too, Montgomery contributec substantial articles for Dionysiu Lardner's *The Cabinet Cyclopaedia* covering the lives and works of Dant Alighieri, Torquato Tasso and Ariosto He also published that year, both i *The Amethyst* and *The Poet's Portfolio* *For ever with the Lord* (see pages 11 and 115) divided in the Portfolio int two parts of nine stanzas and thirtee stanzas each of four lines removin just one verse in his later definitiv text. Selected verses are sung t MONTGOMERY (Woodbury), OLI 50TH, OLD 25TH, WINCHESTE NEW (CRASSELIUS (Hamburg) WER NUR DEN LIEBEN GOTT) and ST PETERSBURG (RUSSIA (Bortnianski); WELLS (Bortnianski) WELLSPRING). An appreciation wa written that year of James Montgomer by Barbara Hofland who had received hi encouragement as Barbara Hoole in he early days as a poet. That year he wa granted a state pension of £200 pe annum by Sir Robert Peel.

By now, most of Montgomery's lif work was complete, but he kept his han in from time to time. In 1836, he wrot an introduction for Bishop Georg Horne's *Commentary on the Psalms*. Hi close friend Samuel Roberts published book on the treatment of gypsies. Jame was invited from time to time to soci occasions. On 18th October 1836, dinner date was suggested with th members of the Montgomery Friendl Society. While in Exeter, he met th Methodist missionary, the Reveren John Beecham who followed up th conversation in a letter in 1837 abou evidence given to the Parliamentar Committee on Aborigines. In 183 Mr. N. C. Brooks solicited a contributio to the *North American* Quarterl magazine published in Baltimore. Ann Gales, with whom he had bee associated from his arrival in Sheffiel lied. He wrote the hymn *One song of* *raise, one voice of prayer* for the Vesleyan Centenary in 1839. He took he New Testament miracles of a leper leansed, the impotent man, the youth orn blind, Lazarus raised from the lead, the withered hand, and the Legion enturion for *Our Saviour's Miracles:* *Sketches in Verse* published in 1840. About this period of his life he undertook several lecture tours. Invitations came from as far afield as Newcastle-under-Lyme, Worcester, Hull and Bath, and nearer home from Bradford. The prospectus arranged by the Bath Royal Literary and Scientific Institution cited forty-three male poets and, unnamed individually, the British poetesses (see Table on page 119). In advertising the programme, the poems *The World before the Flood* and *The Pelican Island* were named as the Montgomery texts most likely to be familiar to the literary public of the day.

Blow ye the trumpet abroad o'er the sea;
Britannia hath conquer'd, the Negro is free:
Sing, for the pride of the tyrant is broken,
His scourges and fetters all clotted with blood,
Are wrench'd from his grasp, for the word was but spoken,
And fetters and scourges were plunged in the flood:
Blow ye the trumpet abroad o'er the sea,
Britannia hath conquer'd, the Negro is free.

James Montgomery: From 'The Negro is Free', No.2 of 'Songs on the Abolition of Negro Slavery in the British Colonies' printed in 'A Poet's Portfolio' with Moore's melody of 'Sound the loud timbrel o'er Egypt's dark sea' in mind

Britain! *not now* I ask of thee
Freedom, the right of bond and free;
Let Mammon hold, while Mammon can,
The bones and blood of living man;
Let tyrants scorn, while tyrants dare,
The shrieks and writhings of despair;
An end *will* come, – it will not wait,
Bands, yokes, and scourges have their date,
Slavery itself must pass away,
And be a tale of yesterday.

James Montgomery: From 'A Cry from South Africa'

In 1841, he mourned the death of his younger brother, Ignatius. He wrote *Within Thy courts have millions met*, much better known in the United States than the United Kingdom. It is sung to LUDBOROUGH. He composed a Jubilee Hymn for Argyle Chapel, Bath, whose minister had served them for fifty years. He accompanied Reverend Peter La Trobe to attend meetings of the Moravian Missionary Society in Edinburgh, Glasgow and Ayr, where public breakfasts were held in his honour. The trip also took in Perth, Dundee and Aberdeen. During this trip he was conscious of Sarah Gales left alone back in Sheffield. Peter La Trobe was also a composer and organist and published *Introduction on the Progress of the Church Psalmody* for an edition of *Moravian Hymn Tunes.* One mundane task that came Montgomery's way was a request from a shareholder to attend the half yearly meeting of the local Gas Company. In those days, attendance by shareholders in person or by proxy was compulsory on pain of forfeiting the shares. James Huie, another of Montgomery's correspondents, published his *History of Christian Missions from the Reformation to the Present Time* and its introduction reminds us of the urgency felt in those missionary days of rescuing the perishing.

My brother Robert reached Ockbrook half an hour afterwards, and for the first time in twenty past years, we *three* met again under the same roof – the only children of parents who had been nearly half a century in their far distant and separate graves – our Mother in Tobago, our Father in Barbadoes. The threefold cord of fraternity was at length broken, the latest spun thread having lasted more than three score years. We, the two remaining survivors, almost immediately on our meeting went into the chamber where our younger Brother's dear remains lay in a coffin . . .

James Montgomery writes from Ockbrook to Sarah Gales on 5th May 184

I

For ever with the Lord!
Amen; so let it be;
Life from the dead is in that word,
'Tis immortality.

Here in the body pent,
Absent from Him I roam,
Yet nightly pitch my moving tent
A day's march nearer home.

My Father's house on high,
Home of my soul, how near,
At times to faith's foreseeing eye
Thy golden gates appear!

Ah! then my spirit faints
To reach the land I love,
The bright inheritance of saints,
Jerusalem above.

Yet clouds will intervene,
And all my prospect flies,
Like Noah's dove, I flit between
Rough seas and stormy skies.

Anon the clouds depart,
The wind and waters cease,
While sweetly o'er my gladden'd heart,
Expands the bow of peace.

Beneath its glowing arch,
Along the hallow'd ground,
I see cherubic armies march,
A camp of fire around.

I hear at morn and even,
At noon and midnight hour,
The choral harmonies of heaven,
Earth's Babel-tongues o'erpower.

Then, then I feel that He
(Remember'd or forgot,)
The Lord, is never far from me,
Though I perceive Him not.

II

In darkness as in light,
Hidden alike from view,
I sleep, I wake, as in *his* sight,
Who looks existence through.

From the dim hour of birth,
Through every changing state
Of mortal pilgrimage on earth,
Till its appointed date.

All that I am, have been,
All that I yet may be,
He sees at once, as He hath seen,
And shall for ever see.

How can I meet His eyes?
Mine on the cross I cast,
And own my life a Saviour's prize,
Mercy from first to last.

For ever with the Lord!
– Father, if 'tis Thy will,
The promise of that faithful word,
Even here to me fulfil.

Be thou at my right hand,
Then can I never fail;
Uphold Thou me, and I shall stand,
Fight, and I must prevail.

So when my latest breath
Shall rend the veil in twain,
By death I shall escape from death,
And life eternal gain.

Knowing as I am known,
How shall I love that word,
And oft repeat before the throne,
For ever with the Lord!

Then, though the soul enjoy
Communion high and sweet,
While worms this body must destroy,
Both shall in glory meet.

The trump of final doom
Will speak the self-same word,
And heaven's voice thunder through the tomb,
For ever with the Lord!

Then, upward as they fly,
That resurrection-word
Shall be their shout of victory,
For ever with the Lord!

The tomb shall echo deep
That death-awakening sound;
The saints shall hear it in their sleep
And answer from the ground.

That resurrection-word,
That shout of victory,
Once more, – *For ever with the Lord!*
Amen; so let it be!

James Montgomery: 'At Home in Heaven' as printed in 'A Poet's Portfolio', 1835

Under the title *China Evangelised: An Ode respectfully addressed to the Christian Missionary Societies of Britain*, he published *Lift up your heads, ye mighty gates* in the *Evangelical Magazine* of 1843. Selected verses are sung as a hymn to PRAETORIUS (FÜR DEIN EMPFANGEN SPEIS UND TRANK), ST DAVID (Ravenscroft), BROMSGROVE (Dyer), CRUCIS VICTORIA, EVANGEL (Fink) and LONDON NEW.

It is to us a subject of deep thanksgiving that many thousands of perishing heathen have been brought to the knowledge and reception of *the truth as it is in Jesus* through the instrumentality of British Christians.

James A. Huie: 'History of Christian Missions from the Reformation to the Present Time'

Rowland Hodgson, George Bennett and Samuel Roberts, friends for a quarter of a century, were still meeting at each others' homes at regular intervals for the purpose of advising on and promoting affairs of benevolence.

In 1844, Isaac Pitman was giving a series of lectures in Sheffield on his new shorthand and the third of these was chaired by James Montgomery. He was appointed the following year as an honorary member of the Sheffield Club and again lectured at the Mechanics' Institute. He wrote an introduction for William Illingworth's *A Voice from the Sanctuary*. He produced a substantial introductory essay for an edition of John Bunyan's classic, *A Pilgrim's Progress*. He provided a hymn for the Jubilee of the Religious Tract Society, beginning *Proclaim the year of Jubilee*.

But I may tell you what you do *not* know, and which it will not be too late for you to learn *on* Monday, – that at 5 minutes past 3 o'clock, at No 56, Pembroke Place, Liverpool, I am thinking of you; – I am with you in spirit; and if our spirits could compare thoughts (meeting at The Mount) yours would tell mine that *it* is just now wondering when we shall meet *bodily* . . .

James Montgomery in a letter to Sarah Gales on 21st August 1844

In 1849 the *Liturgy and Hymns for the use of the Protestant Church of the United Brethren*, published in 1826, was revised under the editorship of James Montgomery and others, with many of the Brethrens' hymns replaced by standard non-Moravian Hymns. He spent a holiday in 1849 in Brighton.

Hugh Miller of the *Edinburgh Witness* wrote of James *there is a green old age in which the spirits retain their buoyancy and the intellect its original vigour.* But he experienced *a slight threatening of paralysis* and from here on, fellow citizens said *How faded! How infirm!*

Part I

Lift up your heads, ye gates of brass!
Ye bars of iron! yield;
And let the King of Glory pass, –
The Cross is in the field.

That banner, brighter than the star,
That leads the train of night,
Shines on their march and guides from far
His servants to the fight.

A holy war those servants wage;
– Mysteriously at strife,
The powers of heaven and hell engage
For more than death or life.

Earth's rankest soil they see outspread;
So throng'd, it seems within,
One city of the living dead,
Dead while alive to sin.

The forms of life are everywhere,
The spirit nowhere found;
Like vapours kindling in the air,
Then sinking in the ground.

No hope have these above the dust,
No being but a breath;
In vanity and lies they trust:
Their very life is death.

Part II

Ye armies of the living God,
His sacramental host!
Where hallow'd footsteps never trod,
Take your appointed post.

Follow the Cross, the ark of peace
Accompany your path,
To slaves and rebels bring release
From bondage and from wrath.

A barley-cake o'erthrew the camp
Of Midian, tent by tent,
Ere morn the trumpet and the lamp
Through all in triumph went.

Though China's sons like Midian's fill
As grasshoppers the vale,
The sword of God and Gideon still
To conquer cannot fail.

As Jericho before the blast
Of sounding rams' horns fell,
Sin's strongholds here shall be down cast,
Down cast these gates of hell.

Truth error's legions must o'erwhelm
And China's thickest wall,
(The wall of darkness round her realm,)
At your loud summons fall.

Though few and small and weak your bands,
Strong in your Captain's strength,
Go to the conquest of all lands,
All must be His at length.

The closest seal'd between the poles
Is open'd to your toils;
Where thrice a hundred million souls
Are offer'd you for spoils.

Those spoils, at his victorious feet,
You shall rejoice to lay,
And lay yourselves, as trophies meant,
In His great Judgement-day.

Part III

No carnal weapons those ye bear,
To lay the aliens low;
Then strike amain, and do not spare,
There's life in every blow.

Life! – more than life on earth can be;
All in this conflict slain
Die but to sin, – eternally
The crown of life to gain.

O fear not, faint not, halt not now;
Quit you like men, be strong;
To Christ shall Buddhu's votaries bow
And sing with you this song:

Uplifted are the gates of brass,
The bars of iron yield;
Behold the King of Glory pass;
The Cross hath won the field.

James Montgomery: 'China Evangelised'

Tracts have the gift of tongues; they preach
Through every peopled land,
In all the forms of human speech,
What all may understand.

James Montgomery: From
'Proclaim the year of Jubilee'

Safe from the world's alluring harms,
Beneath His watchful eye,
Thus in the circle of His arms,
May we for ever lie.

James Montgomery: Verse 5 of
'Children recalling Christ's Example and His Love'
which begins 'When Jesus left his Father's throne'
It is sung to KINGSFOLD

An hundred years ago? – what then?
There rose, the world to bless,
A little band of faithful men,
A cloud of witnesses:–

It look'd but like a human hand;
Few welcomed, and none fear'd,
Yet, as it open'd o'er the land,
The hand of God appear'd.

James Montgomery: Verses from 'A Hymn
for the Wesleyan Centenary' which begins
'One song of praise, one song of prayer'

Fulneck-Hill to-day shall be
Our delightful Bethany;
Dwell, Lord Jesus, where we dwell,
God with us, Immanuel!

James Montgomery: From
'The Christian Sisterhood'
which begins 'On His pilgrimage of woe'

Lord, for ever at Thy side,
Let my place and portion be;
Strip me of the robe of pride,
Clothe me with humility.

James Montgomery: From 'Prayer for Humility'
sung to SONG 13 (CANTERBURY;
SIMPLICITY (Gibbons))

Millions within Thy courts have met,
Millions this day before Thee bow'd;
Their faces Zion-ward were set,
Vows with their lips to Thee they vow'd:

James Montgomery: Opening verse of 'Evening Song for the Sabbath-day'

. . . Mind is invisible; yet when we write,
That world of mystery comes forth to sight;
In vocal speech, the idle air breathes sense,
An empty sound becomes intelligence.
Phonetic Art hath both these modes outdone,
By blending sounds and symbols into one.

James Montgomery, when chairing a lecture about 'Shorthand' at Cutlers' Hall, Sheffield

Forsake him not in his old age,
But while his Master's Cross he bears,
Faith be his staff on pilgrimage,
A crown of glory his grey hairs.

James Montgomery: From 'Prayer for an Aged Minister' beginning 'A blessing on our pastor's head'

In 1852, a full length portrait was painted by Richard Smith. He wrote some verses for the Christian Sisterhood on the centenary of their establishment. On his eightieth birthday, the ladies of Sheffield presented Montgomery with a carved walnut easy chair, plus £50 for the Moravian Fund, £60 for the Aged Female Society, and £60 (including £35 invested to provide a Montgomery Silver Medal each year) for the Sheffield School of Art and Design.

If my name is to be announced in any way do not let the foolish letters *Esq.* be attached to it.

James Montgomery

He checked his texts and issued a definitive edition in *Original Hymns for Christian Worship* in 1853. He died on April 30th, 1854. In his will, he remembered the Moravian Brethren of Fulneck, Moravian Missions, Sheffield Boys Charity School, Sheffield Girls' Charity School, the Society for Bettering the Poor in Sheffield, the Sheffield Aged Female Society, Carver Street National School, the Lancasterian School for Boys, the Lancasterian School for Girls, Sarah Gales, and the editor John Holland, with the residue going to his brother Robert.

Two years later there was concern that the planned public monument had not made its appearance – there was not even a stone above his grave. By then, though, Montgomery was a member of the Church Triumphant and hardly troubled by such earthly honours! But others since have thought to name buildings after him and otherwise show esteem: and a large East window in Sheffield Cathedral was dedicated to his memory.

O Death, how ends thy strife?
– In everlasting life.

James Montgomery: From 'Questions and Answers'

The Poets included in James Montgomery's Lectures
to Bath Royal Literary and Scientific Institution

Joseph Addison (1672 - 1719)
Mark Akenside (1721 - 1770)
Robert Burns (1759 - 1796)
Samuel Butler (1612 - 1680)
George Gordon Byron (1788 - 1824)
*Thomas Campbell (1777 - 1844)
Geoffrey Chaucer (1340 - 1400)
Charles Churchill (1731 - 1764)
Samuel Taylor Coleridge (1772 - 1834)
William Collins (1721 - 1759)
Abraham Cowley (1618 - 1667)
William Cowper (1731 - 1800)
George Crabbe (1754 - 1832)
Richard Crashaw (1612 - 1649)
John Denham (1615 - 1669)
John Donne (1572 - 1631)
John Dryden (1631 - 1700)
Oliver Goldsmith (1731 - 1774)
John Gower (1330 - 1408)
Thomas Gray (1716 - 1776)
George Herbert (1593 - 1633)
Ben Jonson (1573 - 1637)
William Langland (?1330 - ?1400)
John Lydgate (1370 - 1452)
Andrew Marvell (1621 - 1678)
John Milton (1608 - 1674)
*Thomas Moore (1779 - 1852)
Thomas Parnell (1679 - 1718)
Alexander Pope (1688 - 1744)
Matthew Prior (1664 - 1721)
Francis Quarles (1592 - 1644)
Thomas Sackville (1536 - 1608)
Walter Scott (1771 - 1832)
William Shakespeare (1564 - 1616)
John Skelton (1460 - 1529)
Robert Southey (1774 - 1843)
Edmund Spenser (1552 - 1599)
Howard Henry Surrey (1517 - 1547)
James Thomson (1700 - 1748)
Edmund Waller (1606 - 1687)
George Wither (1588 - 1667)
*William Wordsworth (1770 - 1850)
Edward Young (1683 - 1765)
The British Poetesses

* Still living at the time of the lectures

Memorial, Sheffield

Memorial lich-gate, Monkland Parish Church

Henry Baker

If a man gives a lecture, let it be on a subject which he really ought to know something about! Now, I ought to know something about hymns and their writers, not only because it is within the strict and proper range of a parson's learning, but because I have been interested in the subject, more than most for several years past, and for the last year or two have given all the time I could get to the endeavour to make, in union with other clergymen, a collection of hymns which we have some reason to hope will, when published, be less unworthy of use than others have been. And one difficulty we find meeting us in that effort, is just the one I most feel to-night: an *embarras des richesses*, as the French would say.

Our stores of material are so abundant; the mines into which we have to dig, and out of which we have to bring what may be most precious, are so rich and deep; the voices we have to blend together into one outburst of harmony come floating on the breeze from so far distant lands, and so many an age and so various conditions; emperors and kings, bishops and priests, monks in their cloisters, and missionaries on their travels of love, all swelling the strain; it is not easy indeed to know which song is the sweetest, which offering the best.

Oh! what a wonderful uplifting of the human heart to its God has there been, and will be in hymns; from the first song that is on record when the hosts of Israel stood on the shores of the Red Sea, and beneath its waters lay the chariots of Pharoah, his horsemen, and his army, and the chant of Moses and his brethren, *Sing ye to the Lord for He hath triumphed gloriously; the horse and his rider hath He thrown into the sea*, was answered by the glad response of Miriam and her choir of maidens, onwards to that last strain of triumph when they that have gotten the victory for ever, over deadlier enemies than Egypt's king, and through a mightier leader than Moses, shall stand upon the sea of glass mingled with fire, and sing the everlasting song which unites in itself the teaching of the old dispensation and the gospel tidings of the new; *the song of Moses and the song of the Lamb*, as St. John calls it; *Great and marvellous are Thy works Lord God Almighty; just and true are Thy ways Thou King of Saints.*

In that wonderful sacrifice of praise there blend, not unmeetly, Hannah's thanksgiving for Samuel, and David's psalms, and Habakkuk's song of trustfulness, and the *Benedicite of the three children*, with St. Mary's *Magnificat*, and Simeon's *Nunc dimittis*, and the hymn of Zacharias, and the anthem of the four Living Creatures. St. Ambrose and Ephrem Syrus and St. Gregory the Great, and Venantius Fortunatus, and St. Bernard of Clairvaux, and his namesake

Co-proprietor Reverend Christopher Robert Harrison was educated at All Souls' College, Oxford; he served successively as Rector of Leigh, Essex from 1852 to 1855 and of Peldon, Essex from 1855 to 1867; he was Vicar of North Curry, Somerset, from 1867 to 1877. He was one of the earliest members of the Society of the Holy Cross, a congregation of secular priests living under a common rule. It is the oldest catholic society of Anglican priests: Harrison joined the Society in 1856, the year following its foundation by F. C. F. Lowde. He was active in parish work until a few weeks before his death. In his last year he set in motion an appeal for £4,000 to restore the church then *in a condition of great dilapidation and decay*. A report in *The Guardian*, mentioning his death at Builth Wells where his brother was the incumbent, refers to him *as a man of great judgement, earnestness, ability and true devotion* who had contributed letters to the paper to call *attention to the dangers and difficulties attending the cry for the alterations to the laws of burial.*

The great West Window of North Curry Church was installed in memory of Harrison. It represents Dean Alford's hymn *Ten thousand times ten thousand* as an allusion to Harrison's work on *Hymns Ancient and Modern*. The window contains his portrait and that of his son who fell in battle in 1881.

Co-proprietor Reverend George William Huntingford was a Fellow of New College, Oxford and served as an assistant master of Winchester College. In 1851, he became Vicar of Littlemore. In the eighteenth century, the Oxfordshire village suffered pastoral neglect: but when Reverend J. H. Newman (later Cardinal Newman after his move to the Roman Catholic Church) was presented in 1828 to the living of St. Mary's, Oxford, he had charge also of Littlemore. Against much opposition, he achieved its establishment as a separate parish. The foundation stone of the church was laid in 1835; its furnishings were largely planned by Reverend J. R. Bloxam, the father of all ritualists, who became curate in 1837. The work of enlarging the church was completed in 1848: the tradition inherited by Huntingford is self-evident. In 1872, Huntingford moved to the quiet village of Barnwell near Oundle. A window to the memory of George Huntingford and his wife is underlaid with a quotation by the seventeenth century poet Henry Vaughan.

> **God's saints are shining lights**
> **Who stays here long, must passe**
> **O'er dark hills, swift streams, steep ways**
> **As smooth as glasse**
> **But these all night like candles shed**
> **Their beams, and light us unto bed**

He came to a church with a beautiful east window installed in 1851 and designed by Sir Giles Gilbert Scott. An organ was installed in 1874, two years after Huntingford's arrival.

Henry Baker

of Cluny, and Adam of St. Victor, and Thomas de Celanc and St. Francis Xavier swell the chorus, with our ow Venerable Bede and Sedulius, and saintly Ken, an missionary-bishop Heber, and George Herbert, and Joh Keble; yes, and Charles Wesley and Augustus Toplady; fo the *Jesu, lover of my soul* and *Rock of Ages* are surely no unworthy of a place amongst the Church's songs.

These words are reprinted from *The Hereford Journal* o 7th March, 1860 as they report a lecture given by Reveren Sir Henry Baker, Bart. at Leominster Town Hall. Baker wa born in London on 27th May, 1821, eldest son of Vice Admiral Sir Henry Loraine Baker, second baronet, an Louisa Anne, the only daughter of William Williams, some time Member of Parliament for Weymouth. He was educate at Trinity College, Cambridge, graduating Bachelor of Arts i 1844 and Master of Arts in 1847. After ordination, he serve as curate at Great Horkesley, a village a few miles to th north of Colchester. Here he fell in love with Letitia Bonno but his rival, the Reverend W. Edwards, who had also been curate at Great Horkesley, won her hand. Edwards becam Vicar of Orleton: whilst in 1851, Baker was similarl appointed to Monkland, both near to the English marke town of Leominster. The East windows of Orleton Churc are filled with stained glass in memory of Letitia: Edward nephew wrote that from the time Letitia rejected Henr Baker *there was no further friendly communication betwee the vicarages of Orleton and Monkland*. Baker remained bachelor and later was well known as an advocate of clerica celibacy.

The obscure village of Monkland on the River Arrow was home to rather more than 200 souls. When the incumbent arrived, there was no vicarage: consequently, he had Horkesley House built, named after the place of his curacy. The design included a private chapel with an organ incorporated. The house still stands, though divided into three. Baker's love of sacred music meant that any applicant to work in his household, indoor or outdoor, needed an additional talent – an ability to sing in the choir. During his incumbency he had two cottages built especially for the use of organist and choirmaster. Soon after his arrival at Monkland, he wrote his first hymn, *Oh, what if we are Christ's* written initially for St. James' Day and usually sung to ST. MICHAEL (OLD 134th). It was published by Reverend Francis H. Murray, the Rector of Chislehurst from 1843 to 1902 in his 1852 *Hymnal for use in the English Church.*

Oh! what, if we are Christ's,
Is earthly shame or loss?
Bright shall the crown of glory be,
When we have borne the cross.

Keen was the trial once,
Bitter the cup of woe,
When martyred Saints, baptized in blood,
Christ's sufferings shared below;

Bright is their glory now,
Boundless their joy above,
Where, on the bosom of their God,
They rest in perfect love.

Lord! may that grace be ours,
Ever like them to bear
All that of sorrow, grief, or pain
May be our portion here;

Enough if Thou at last
The word of blessing give;
And let us rest beneath Thy Feet,
Where Saints and Angels live.

Give to the Father praise,
Praise to the Holy Son,
And praise the Holy Spirit's Name,
Eternal Three in One.

As printed in 'A Hymnal for use in the English Church' (1852) edited F. H. Murray beneath the text 'Ye shall indeed drink of My cup, and be baptized with the baptism that I am baptized with'. The tune recommended was LEEDS

Co-proprietor Reverend John Murray Wilkins was educated at Trinity College, Cambridge. He served first as Curate of St. Mary's, Nottingham, but from 1840 his life was spent as Rector of the Collegiate Church of Southwell, at a period when there was considerable tension as powers of administration were passing from the ancient Chapter to central authorities.

The collegiate church constitution had remained unchanged from the days of King Edgar until Victorian times. The chapter house earned the reputation of being perhaps the most beautiful gem of Gothic architecture in the world. In 1884, after Wilkins' death, it was to become a Cathedral.

'Observe the very large share that the Church of England, according to the Bible pattern, assigns and prescribes to her *people* in the celebration of the public offices. Her ritual, her liturgy, her worship, is not to be a chill, cold dialogue between the minister and the clerk, with the addition now and then of the feeble voices of a few children or choristers, of which the people are to be mere spectators, or audience, or critics, or silent, or even only murmuring, partakers . . .'

Sermon preached by Rev. John Murray Wilkins on 3rd May, 1860, at the third annual festival of Parochial Choirs of the Nottinghamshire Church Choral Union.

In 1851, the children of four labourers, the publican and the carpenter were baptized in Monkland.

The architect of the vicarage, then little known, was George Edmund Street. In his later career, he was to build the new Law Courts in London, to restore the cathedrals of York, Salisbury and Carlisle in England and St. Brigid's at Kildare in Ireland; the chapel of Haddo House, home of the Gordons, and many parish churches. He was in great sympathy with the Tractarian movement and built for the Liturgy. As an architect he believed he should himself be accomplished in all crafts: he taught himself joinery and carpentry, the art of designing and painting stained glass, and the work of a blacksmith. He also gained a reputation of giving the same attention to detail to the smaller of jobs as to larger ones.

The former Vicarage, Monkland

The village of Monkland had no school; so Baker arranged in 1852 for a parcel of glebe land belonging to the Vicarage to be conveyed as a site for this purpose. The school provided places for sixty children. It is used today as a Village Hall. His hymn for use at school festivals, *Lord Jesus, God and Man*, sung to CARING or ST. HELENA, was written the same year. In 1853, a rate of fourpence in the £ was levied for repairs to the church and in 1859, two cracked bells were sent for repair.

The arrival of the railway at Leominster made Monkland easily accessible from other parts of the country: and it was to become the centre of Baker's life-work. For the story of Sir Henry is more than that of a hymnwriter: it is the story of a famous book, *Hymns Ancient and Modern*. To understand how that book began, it is necessary to go back a further three hundred years. In 1562, the metrical psalms of Sternhold and Hopkins were published as *The Whole Psalms*, commonly known as the *Old Version*. In 1696 the *New Version*, by Nicholas Brady and Nahum Tate, was authorised as an alternative. This also included the *Veni Creator*, the *Ten Commandments*, the *Athanasian Creed*, the *Twelve Articles of Christian Faith*, the *Venite*, four *Gloria Patri* and seven prayers and devotions. In the Church of England, almost every congregation was still using either the *Old Version* or the *New Version* at the beginning of the nineteenth century as the only hymns in use. Whilst many chapels rang with the new hymns of faith of the nonconformists, most Anglicans had to be content with a principal diet of the pre-Christian Psalms.

Lord Jesus, God and Man,
For love of man a Child,
The Very God, yet born on earth
Of Mary undefiled;

Lord Jesus, God and Man,
In this our festal day
To Thee for precious gifts of grace
Thy ransom'd people pray.

We pray for childlike hearts,
For gentleness and love,
For strength to do Thy will below
As Angels do above.

We pray for simple faith,
For hope that never faints,
For true communion evermore
With all Thy blessèd Saints.

On friends around us here
O let Thy blessing fall;
We pray for grace to love them well,
But Thee beyond them all.

O joy to live for Thee!
O joy in Thee to die!
O very joy of joys to see
Thy face eternally!

Lord Jesus, God and Man,
We praise Thee and adore,
Who art with God the Father One
And Spirit evermore.

Henry Baker

Especially from 1820 onwards, many parish incumbents introduced their own books. That year, Reginald Heber, later Bishop of Calcutta, submitted his parish hymn book for use at Hodnet in Shropshire to the Archbishop of Canterbury for approval which was refused: Thomas Cotterill, Vicar of St. Paul's, Sheffield, secured episcopal approval for a new edition of his *Selection of Psalms and Hymns*.

Co-proprietor Reverend Francis Henry Murray was educated at Christ Church, Oxford: he served as Curate of Northfield, Birmingham from 1843 to 1845 and as Rector of Chislehurst, Kent, from 1846 to 1902; he was an Honorary Canon of Canterbury. In 1852, he edited ***A Hymnal for use in the English Church*****, also known as** ***Mozley's Hymnal.*** **In the Preface, the purpose is stated 'to provide a Hymnal, which might be, as far as they could make it, both in doctrine and arrangement, a companion and complement to the Book of Common Prayer'. Accordingly the basis of the work, like that of the Prayer-Book, was assumed to be the ancient Hymns of the Church, which, like those Prayers, have been the common inheritance of Christians from the earliest times. There could be, for instance, no question about the adoption of Hymns composed as early as, even if St. Ambrose was not himself the author of, the** ***Te Deum.*** **And, as in the Book of Common Prayer there are compositions of a later date, so it was judged that Hymns also of more modern composition might be fully admissible. For as, on the one hand, the Church has signified her approval of the Ancient Hymns in the use of** ***Veni Creator*** **in the Ordination Service; so, on the other, a sort of** ***quasi*** **authority and guide is given for the adoption of modern Hymns at the end of the Prayer-Book. And this is nothing more than the prerogative and power of the Church thus to take up, and employ in her own use** ***things new and old.***

Monkland School - now the Village Hall

Towards the end of the 1830's, the Tractarians looked to the hymns of the Early Church for material and in 1837 Bishop Mant published *Hymns from the Roman Breviary for Domestic Use* and Isaac Williams published *Hymns from the Parisian Breviary*. These were intended for private devotions but in 1841, John Chandler re-issued some of his translations in a form designed for public worship as *The Hymns of the Church, mostly Primitive*. In 1852 and 1854, a group led by J. M. Neale published the *Hymnal Noted*, wholly consisting of versions of Latin hymns and probably the most important such work. Neale was to go on to translate from the Greek the Hymns of the Eastern Church. While all this work was proceeding, there was a bonanza of new texts being published by contemporary writers. The musical choice was as broad as the literary. Sir John Goss had in 1827 issued the first part of *Parochial Psalmody* for use by Chelsea Parish Church. He called for a recovery of treasures from the past, whilst Dr. Crotch wrote very clearly on the need to ban unworthy tunes in his preface to *Psalm Tunes for Cathedrals and Parish Churches* in 1836. In 1847 Havergal, through his *Old Church Psalmody* brought many German chorales into use. Dr. Henry Gauntlett and the Reverend W. J. Blew published the *Church Hymn and Tune Book* in 1852, which brought together the new kind of hymn tune born of the nineteenth century. There was a bonanza of locally issued tune books to match the bonanza of new words. Quantity did not necessarily mean quality!

Co-proprietor Reverend Thomas Astley Maberley was educated at Christ Church, Oxford and served as Curate of St. Andrew's Church, Holborn, London, from 1836 until 1841 and as Vicar of Cuckfield, Sussex, from 1841 to 1877.

Henry Baker was an enthusiastic convert to the Oxford movement. He believed the Church of England needed a hymn-book to accompany its liturgy. Cranmer, who had translated its liturgy and created the Prayer Book, had omitted to supply the hymns with which the ancient service books had been liberally sprinkled. Henry Baker dreamed of this hymn-book as a living reality: though he knew that to gain wide acceptance, it would need to contain modern hymns alongside those inherited from the ancient church.

'The evils of pews . . . in the first place, all unnecessary separation from one another in the house of God is to be condemned, it tends to make distinctions where there ought to be none, and to prevent us from realising properly in our minds that we are offering up our prayers as children of one father, all equal in His sight . . . '
(To his congregation)

Meantime, in 1853, the Rev. G. Cosby White had compiled *Hymns and Introits* for use of the Collegiate Church at Cumbrae in Scotland where he was Provost. That same year he moved to Chislehurst as curate to Rev. Francis Murray who had published Baker's first text. Also in 1853, the Rev. W. Denton edited a successful book entitled *Church Hymnal*. In 1857, in a carriage of the Great Western Railway, Denton suggested to Murray that the three books should be amalgamated. Murray was aware that the Vicar of Monkland was also looking for ways of amalgamating competing books, and contact was made with Baker. He was sympathetic to working together and named Rev. P. Ward, active in the Salisbury Diocese, as another person considering possibilities for amalgamation.

These contacts led to an historic meeting in the Clergy House of St. Barnabas, Pimlico, where Cosby White had recently been appointed Priest-in-Charge. Besides the Chislehurst trio of Murray, Harrison and White, Baker brought his friend, Rev. W. Pulling, Rector of the Herefordshire parish of Eastnor. Somehow Denton was not asked. They decided to go ahead with a new book: Baker was appointed as secretary to the committee. Further persons were co-opted including Rev. J. R. Woodford, afterwards Bishop of Ely, the editor of *Hymns for the Sundays and Holy Days of the Church of England* with editions published in 1852 and 1855; the Rev. W. U. Richards, Priest-in-Charge of Margaret Chapel in London, later first vicar when it became All Saints, Margaret Street; and Rev. G. W. Huntingford, who had served as an assistant master of Winchester College and was Vicar of Littlemore, near Oxford. Henry Baker consulted with many other editors of local hymnbooks and by 27th October, 1858, the group was ready to advertise its plans in *The Guardian*. There were over two hundred replies, including one from William Walsham How, author of *For all the saints who from their labours rest* and another from J. M. Neale, translator of *All glory, laud and honour*.

Co-proprietor Reverend William Pulling, who was sometime Chairman of the Proprietors, was educated at Oriel College, Oxford and served as Fellow and Tutor of Brasenose College from 1839 to 1849; then until 1894, he served as Rector of Eastnor, Herefordshire, also acting from 1850 as Rector of nearby Pixley. The parishes were not far from Sir Henry Baker's home.

Addressing the 1879 Church Congress in Swansea, he reported: The circulation of *Hymns Ancient and Modern* has exceeded 20,000,000 copies, of which no inconsiderable proportion has been given to poor parishes in every part of the United Kingdom, and throughout the Colonies by free grants . . .

The Clergy House and St. Barnabas, Pimlico

Baker was far from being just a committee secretary: he was the driving force behind the whole project. A letter of his in December 1858 gave his views on what should be included: he also believed all hymns chosen should be sent to all correspondents. By indicating that some hymns were not necessarily intended for church worship, he showed the political skill with which he was so often to steer *Hymns Ancient and Modern* into a form acceptable to those of varying churchmanship. Nevertheless, he was not able to hold together all members of the large Planning Committee which was formed in January 1859 and met under Sir Henry's chairmanship. Woodford withdrew on the quality of many of the translations – *mostly laboured renderings, without glow.* The Rev. Francis Pott, who eventually issued his own *Hymns Fitted to the Order of Common Prayer* in 1861 resigned because he disapproved of the doctrinal teaching in some of the Latin hymns.

If you desire to make a hymn-book for the use of the Church, make it comprehensive.

Advice to the Committee from John Keble, author of 'New every morning is the love'

Sir Henry himself made several translations from the Latin and the list of them shows a sprinkling of the sources of *Hymns Ancient* both in terms of words and music which were included in the Original Edition. For Sunday he translated *On this day, the first of days* from *Le Mans Breviary* of 1748 where it stands for use on summer Sunday mornings: it was set to LÜBECK (GOTT SEI DANK), based on a melody in *Geistreiches Gesanbuch*, published in Halle in 1704. For Christmas, he provided *O Christ, Redeemer of our race,* which had remained in use until the eleventh century; it was printed with a plainsong tune. The seventeenth century Parisian writer Jean Baptiste de Santeuil provided the source for *Now, my soul, thy voice upraising*, a Passiontide hymn set to a melody traditionally sung at the Veneration of the Cross on Good Friday. The same author prompted *Captains of the saintly band* to celebrate the life of Apostles on Saints Days. For *Jesu, grant me this, I pray*, he turned to an anonymous text in *Symphonia Sirenum* published in Cologne in 1695, with a tune by Orlando Gibbons written for Wither's *Hymns and Songs of the Church*, in 1623. He also translated *O sacred head, surrounded*, a hymn which comes to us from the Latin through the German of the Lutheran Paul Gerhardt. The Latin poem is the last of a series of seven poems addressed to the several members of Christ hanging on the cross. The tune PASSION CHORALE (O HAUPT VOLL BLUT UND WUNDER; HERZLICH THUT MICH VERLANGEN) was written originally for secular words. He turned to Dean William Bullock, who was Rector of the Church in Trinity Bay, Newfoundland for *We love the place, O God.* The rector possibly wrote it for the consecration of the church there in 1827: Baker adapted the first four verses

. . . *all* hymns we take from the *Sarum* Breviary (and perhaps Roman also) be translated in the metres of the original; and if they are given in the *Hymnal Noted* that they be made to suit the tunes there; but that with regard to hymns from other ancient sources, the *Paris*, e.g., we be at liberty to do as we like . . .

. . . to *profess* to give among the general hymns some that would be suitable for singing in mission rooms, at lectures in cottages, etc., rather than in church. We want such – and this would give *good grounds* on which to admit some as to which we should otherwise feel bound to be more strict.

Extracts from a letter written by Sir Henry Baker on 15th December 1858

Permission was also granted to the Vicar to send the 2 cracked bells to Messrs. Warner & Sons Bellfounders in London to be recast – the cost of the same to be defrayed by a Voluntary subscription and the Vicar to be responsible for their return.

Minutes of Monkland Vestry: 20th August 1859

and added the rest. It was originally set to QUAM DILECTA by Rev. Henry Jenner, a fellow-student of Baker's at Trinity College, Cambridge and later Bishop of Dunedin. It is also sung to ANNUE CHRISTE (LA FEILLÉE).

We love the place, O God,
Wherein Thine honour dwells;
The joy of Thine abode
All earthly joys excels.

We love the house of prayer,
Wherein Thy servants meet;
And Thou, O Lord, art there
Thy chosen flock to greet.

We love the sacred font;
For there the Holy Dove
To pour is ever wont
His blessing from above.

We love Thine altar, Lord:
Oh, what on earth so dear?
For there, in faith adored,
We find Thy presence near.

We love Thy saints who come
Thy mercy to proclaim,
To call the wanderers home,
And magnify Thy name.

Our first and latest love
To Zion shall be given,
The house of God above,
On earth the gate of heaven.

We love the word of life,
The word that tells of peace,
Of comfort in the strife,
And joys that never cease.

We love to sing below
For mercies freely given;
But, Oh, we long to know
The triumph-song of heaven.

Lord Jesus, give us grace
On earth to love Thee more,
In heav'n to see Thy face,
And with Thy Saints adore.

Henry Baker

Captains of the saintly band,
Lights who lighten every land,
Princes who with Jesus dwell,
Judges of His Israel.

On the nations sunk in night
Ye have shed the Gospel light;
Falsehood flies before the day;
Truth is shining on our way.

Not by warrior's spear and sword,
Not by art of human word,
Preaching but the Cross of shame,
Rebel hearts for Christ ye tame.

Earth, that long in sin and pain
Groan'd in Satan's deadly chain,
Now to serve its God is free
In the law of liberty.

Distant lands with one acclaim
Tell the honour of your name,
Who, wherever man has trod,
Teach the mysteries of God.

Glory to the Three in One
While eternal ages run,
Who from deepest shades of night
Call'd us to His glorious light.

This hymn has been sung to VIENNA (Pleyel) and UNIVERSITY COLLEGE

By May 1859, a specimen booklet of fifty *Ancient and Modern* texts was sent out, and on 18th November of that year, a little book of 138 hymns was issued. Just how the larger committee became the eleven Proprietors who financed the enterprise is unrecorded. Besides the five who had first met at St. Barnabas Clergy House, they were Huntingford and Richards, the Rev. W. H. Lyall, Rector of St. Dionis Backchurch in the City of London, Rev. T. A. Maberley, the Vicar of Cuckfield in Sussex, the Rev. J. Wilkins, Rector of Southwell Minster and Rev. P. Ward, who was to resign in 1863, feeling bound to help the *Salisbury Hymnal* being produced in his Diocese. Baker presided over a diverse group. The Rev. W. U. Richards, for instance, wanted *Ancient and Modern* to publish two hymns for each day of the week for Matins and Evensong, and hymns for all the Festivals including those of the minor Black Letter Saints. In the end, not even all the Red Letter Saints were covered. Baker's tremendous energy ensured publication of the Words Edition of 273 hymns by Advent Sunday 1860, with the Music Edition following in March 1861.

P. Ward, the only short-term co-proprietor, resigned in 1863 feeling bound to help work on the revision of a book for use in the Diocese of Salisbury. Attempts to avoid a publication separate from 'Hymns Ancient & Modern' failed; and *The Sarum Hymnbook* was published in 1868.

During 1860, Guiseppe Gariboldi led his thousand Redshirts to conquer Sicily and Naples for the new Kingdom of Italy; war raged between Spain and Morocco. The French sent an expedition to Syria to protect the Maronite Christians against the Druzes; and intervened in the Lebanon as eleven thousand Christians were massacred by the Ottomans. King Victor Emmanuel invaded the Papal States; Europeans and Maoris fought in New Zealand. There was civil war in Columbia and Mexico . . . Baker contributed *O God of love, O King of peace* which is sung to CANNONS (CANONS), HOLLAND, PHILADELPHIA, ROCKINGHAM (CATON; COMMUNION (Miller)), ERHALT UNS, HERR and ST. GREGORY.

O God of love, O King of peace,
Make wars throughout the world to cease;
The wrath of sinful man restrain,
Give peace, O God, give peace again.

Remember, Lord, Thy works of old,
The wonders that our fathers told,
Remember not our sin's dark stain,
Give peace, O God, give peace again.

Whom shall we trust but Thee, O Lord?
Where rest but on Thy faithful word?
None ever call'd on Thee in vain,
Give peace, O God, give peace again.

Where Saints and Angels dwell above,
All hearts are knit in holy love;
O bind us in that heav'nly chain.
Give peace, O God, give peace again.

Henry Baker

Jesu! Jesu!
By Thy Blood for sinners flowing,
By Thy Death true life bestowing,
We beseech Thee, we beseech Thee,
From every ill defend us.
Thy grace and mercy send us.

Jesu! Jesu!
By Thy glorious Resurrection
Earnest of our own perfection,
We beseech Thee, we beseech Thee.
From every ill defend us,
Thy grace and mercy send us.

Henry Baker: Verses from 'God the Father, from Thy throne' (Litany of Intercession)

Co-proprietor Reverend William Upton Richards was educated at Exeter College, Oxford. He combined duties in the Manuscript Department of the British Museum with the role of Minister at London's Margaret Chapel. Between 1839 and 1845, he was assistant to Reverend Frederick Oakeley, translator of *Adeste, fideles (O come all ye faithful).* **Oakeley described the chapel as a complete paragon of ugliness . . . it bore no resemblance to the Christian fold than that of being choked with sheep-pens under the name of pews . . . In this unpromising place, Oakeley pioneered the working out of Tractarian ideals. When he joined the Church of Rome in 1845, Richards, the cheerful Cornishman, was appointed his successor. His life's work was at Margaret Chapel and in the famous successor-church of All Saints', Margaret Street. He had no wish to follow Oakeley to Rome but quietly pursued the traditions of worship inspired by the Oxford Movement. There were recurrent financial crises in the life of the church and its school: and many Anglicans were suspicious of what went on in worship. His contemporaries judged him a man of great simplicity and great sincerity, piety, gentleness and dogged persistence. He played an important role in the founding of the Society of All Saints (Sisters of the Poor), the first religious order for women in the Church of England since the Reformation. It was Richards who in 1865 proposed the motion in the Proprietors meeting to grant £100 to fill the East Window of Monkland Church with stained glass. John Keble, author of** *There is a book, who runs may read*, **was a close friend and often stayed with Richards. His death in 1866 was the cause of much grief. Richards was the first of the Proprietors to die in 1873.**

For the joyful occasion of Holy Matrimony he found inspiration from the Wedding Feast at Cana, *How welcome was the call.* The tune ST. GEORGE (ST. OLAVE) chosen, was by H. J. Gauntlett, for many years the organist at Olney, where Cowper and Newton had earlier written *Olney Hymns.* The variety of content is further shown with Baker's Litany of Intercession *God the Father, from Thy throne* of which two verses are quoted to show the style. *Hymns Ancient and Modern*, to gain acceptance, ranged wide in subject. With little state help in times of trouble, Benefit Societies helped members through troubled times and Baker's *O Praise our God to-day* sung to BEN RHYDDING or SWABIA celebrates them. For Harvest, he was inspired by John Milton's version of Psalm 136 *Let us with a gladsome mind* to write *Praise, O praise our God and King* which continues to be sung round the world to-day to the tune named after the village of MONKLAND. Equally well-known is *Lord, thy word abideth* to RAVENSHAW, an adaptation made by Monk from an old German melody from the fifteenth century. These were times when the frailty of the human body to withstand sickness was a common experience in most homes; and when dire poverty was all too frequent. As spirituals have looked beyond the slave's lot to the blessed country beyond death, so too did many Victorian hymns. Baker's contribution was *There is a blessed home*, sung to ANNUE CHRISTE (LA FEILLÉE) and UXBRIDGE (HYMN OF EVE).

Baker, the committee chairman; Baker, the business man; Baker, the salesman; Baker, the musician; Baker, the copyright man; Baker, the reconciler; Baker, the persuader all find a place in the ensuing story of the new hymn-book and his Monkland home was often full of visitors. In all this, he enjoyed the help and support of his sister, Jessy, who acted as hostess at Horkesley during her brother's ministry. She remained at Monkland after her brother's death, continuing to support the Church, which continued a rich musical tradition. In 1890, when Jessy Baker eventually moved, the Rev. John Padfield asked *Hymns Ancient and Modern* for financial help in maintaining that tradition.

He had been one of the preachers in a series of sermons at St. Barnabas, Pimlico, the cradle of *Hymns Ancient & Modern*, during the Octave of its Consecration in 1850. His chosen subject was *The Danger of Riches*. 'My brethren, it is very miserable indeed to think that amidst much real improvement of late years amongst ourselves, luxury too has been much increased, and is even now increasing. Miserable, I say, for what is luxury but selfishness, the path to forgetfulness of God, the special hardener of the heart, and the minister to other sins.'

How welcome was the call
And sweet the festal lay,
When Jesus deign'd in Cana's hall
To bless the marriage day!

And happy was the bride,
And glad the bridegroom's heart,
For He who tarried at their side
Bade grief and ill depart.

His gracious power divine
The water vessels knew;
And plenteous was the mystic wine
The wondering servants drew.

O Lord of life and love,
Come Thou again to-day;
And bring a blessing from above
That ne'er shall pass away.

O bless now, as of old,
The bridegroom and the bride;
Bless with the holier stream that flow'd
Forth from thy pierced side.

Before thine altar-throne
This mercy we implore:
As Thou dost knit them, Lord, in one,
So bless them evermore.

Henry Baker

There is a blessèd home
Beyond this land of woe,
Where trials never come,
Nor tears of sorrow flow;
Where faith is lost in sight,
And patient hope is crown'd,
And everlasting light
Its glory throws around.

Henry Baker

The Church was beautifully restored by Sir Henry Baker & he also put a very good organ in it; the living is of the smallest & we have no parishioner above a tenant farmer (except Miss Baker) so that it is impossible to keep up the services of the Church in their present condition without outside help.

From a letter to the Proprietors of 'Hymns Ancient and Modern' from Rev. John Padfield, Vicar of Monkland, 1890

The Book was an immediate success beyond the Proprietors' wildest dreams. At their regular meetings at St. Barnabas' Parsonage in London's Pimlico, they voted some of the proceeds to assist poorer parishes in introducing the Book. They tried hard, when the Salisbury diocese was planning its own book, to establish co-operation, but after lengthy exchanges of view, that Diocese decided to go its own way. The Book was sent round the world as copies went to all the Colonial Bishops: but the diverse concerns of the committee are shown as this promotional exercise shared a committee agenda with 'Insurance of Stocks'. By 1864, two major issues were on the table. In starting the venture, little thought was given to the long term: now they were concerned about the book beyond their own lifetimes. So the Minutes of the meeting of 12th April mention the subject of a Deed of Partnership as well as this resolution of more immediate moment proposed from the Chair:

> . . . That the members of the Committee be requested to write to certain persons interested in hymnody to ascertain their opinion
>
> 1 As to the desirability of publishing an Appendix, it being distinctly understood that the book will be always kept on sale in its present form.
>
> 2 As to any materials for such an Appendix.

The motion was carried unanimously and amongst the certain persons were Archbishop Trench, Brother Morrell, the Reverends Littledale, Keble, Wordsworth, Webb, Neale and Sir John Stainer.

We the undersigned being the major part of the inhabitants and occupiers addressed to the Relief of the Poor of the Parish of Monkland in vestry assembled do hereby resolve and determine that the sum of money which is requisite for the purpose of repairing the walls and roof and floor of the Parish Church amounts to £400 – And we the undersigned do hereby resolve direct and consent that the Wardens and Overseers of the Poor of the said Parish do with the consent of the Bishop of the Diocese and the Incumbent of the said Parish make application to the said Commissioners for a loan of £Two hundred and fifty pounds being part of the sum of money requisite for the purpose aforesaid . . .

Minutes of Monkland Vestry Meeting: 13th October 1864

Monkland Parish Church

\t the committee meeting on 2nd June, ›ffprints were approved from the ›riginal book for numerous different ›ccasions – for Confirmations, Harvest Festivals, Consecration of Churches, Missionary meetings, School Festivals nd Friendly Societies.

All this work on hymnody did not mean ack of attention to Monkland Parish Church. It was in great need of enovation and again, he called in George Edmund Street. Part of the cost was overed by a loan from the Public Works oan Commissioners against Baker's ersonal guarantee. The Proprietors of *Iymns Ancient and Modern* voted £100 or filling the East window with stained lass. At the church re-opening in 1866, n event carried out with great olemnity, the hymn *Lord Thy word bideth* was sung. The Minutes show ther grants for works in several of the roprietors' churches.

His arm the strength imparts
Our daily toil to bear;
His grace alone inspires our hearts
Each other's load to share.

. . . . .

O happiest work below,
Earnest of joy above,
To sweeten many a cup of woe
By deeds of holy love!

. . . . .

Lord, may it be our choice
Thy blessed rule to keep,
'Rejoice with them that do rejoice
And weep with them that weep.'

Henry Baker: Three Verses from 'A Hymn for Benefit Societies' which begins 'O praise our God to-day'

Baker had a difficult meeting with the Dean of Canterbury vho objected that his hymn had been printed with nauthorised changes. It was suggested afterwards in Committee that in the future editions there should be rinted: *The Compilers feel it due to the Very Reverend the Dean to state that considerable alterations were made by hem in this hymn, which although not approving, he most kindly permits them to retain.* Jealous of their copyright, the Proprietors had Baker take the Chaplain-General to task for roducing an edition of the Hymn Book, bearing the title *Iymns Ancient and Modern*, and printed without permission y the King's Printer. A book for Naval use had earlier been roperly authorised.

Amongst the hymns contributed to the Appendix published n 1868 was Sir Henry's text composed in honour of the Blessed Virgin Mary, *Shall we not love thee, Mother dear*, ung to BELMONT. A good deal of anger was expressed in nany quarters over the inclusion of this hymn in the Appendix. To venerate the Virgin Mary was highly bjectionable to many Anglicans. A copy of *Hymns Ancient nd Modern* used in the nearby parish of Pudleton, shows ffensive phrases removed from certain texts, for example n John Keble's *The voice that breathed o'er Eden, Be present, Son of Mary* is crossed through and replaced by *Be present, Holy Jesus*. The chorus to Henry Milman's *When our heads are bow'd with woe* is also crossed through with *Jesus, son of Mary, hear* replaced by *Jesus son of God, give ear*.

Baker also contributed to the Appendix perhaps the best known of all his texts *The King of love my shepherd is*. It is in part a re-cast of George Herbert's paraphrase of the *23rd Psalm*. Its tune DOMINUS REGIS ME was especially composed for the text by Reverend John Bacchus Dykes a parish vicar in Durham, who was consulted widely by Baker in musical matters. Indeed Music Editor Monk clearly did not have a free hand. Besides Dykes, Baker consulted Sir Frederick Gore Ouseley, Bart. who built St Michael's College, Tenbury, which he generously endowed and served as Warden: here chosen musical boys received an education to fit them as leaders in the maintenance of the best possible standards of music in the Church of England. He also consulted John Stainer (of Stainer's *Crucifixion*) both in connection with the Book and his own tunes. At the time Stainer was organist of St. Paul's Cathedral, London. In 1888, he was knighted by Queen Victoria. Baker had the good publisher's instinct for suitable, singable tunes which the Proprietors often bought outright for a few guineas, afterwards guarding their copyright as a valuable commodity. Baker, the business man, was not sure either that the deal they had from Novello & Co. was as good as it should have been and there were lengthy negotiations over costs. In 1870, the Proprietors allowed the tunes to be embossed in Lucan characters by the London Society for the Blind.

The remuneration of the Music Editor features regularly in the Minutes of the Proprietors. Sir Henry and Monk worked closely over the years and the musician was an occasional visitor to Monkland Vicarage. When Monk played the church

Shall we not love thee, Mother dear,
Whom Jesus loves so well,
And to His glory, year by year,
Thy joy and honour tell?

. . . . .

The Babe He lay upon thy breast,
To thee He turn'd for food;
Thy gentle nursing soothed to rest
Th' Incarnate Son of God.

. . . . .

Joy to be Mother of the Lord –
And thine the truer bliss,
In every thought, and deed, and word
To be for ever His.

Henry Baker: Selected Verses of 'Shall we not love thee, Mother dear'

And some within Thy sacred fold
To holy things are dead and cold,
And waste the precious hours of life
In selfish ease, or toil, or strife.

. . . . .

Out of the deep of fear,
And dread of coming shame,
All night till morning watch is near
I plead the Precious Name.

Henry Baker: Verses from two other Baker hymns in the Appendix beginning 'Almighty God, whose only Son' and 'Out of the deep I call'

organ – the best in Herefordshire at that time save for the instrument in the cathedral, the organ blower complained of the hard work he had because Monk *used such a lot of wind.* He does not however appear to have put much worth on Baker's sermons, for when not acting as organist, he is said to have sat in the tower at the West End of the Church and snored audibly. So it was said that he made a great noise at whichever end of the Church he happened to be. When the Appendix to the Original Edition of *Hymns Ancient and Modern* was in preparation, Monk clearly became so exasperated with the tardiness of the Proprietors in paying for his services that he wrote formally to his friend. The letter produced results. Monk's proposals were accepted on condition that his copyrights were formally made over to the Proprietors. Monk and Baker also worked together over many years in producing *The Psalter pointed and set to accompanying Chants 'Ancient and Modern'.* This work was not finally published until after Sir Henry's death.

Sir Henry Baker entered upon the work with the same energy and unwearied direction which he had given in the preparation of *Hymns Ancient and Modern* and it was almost his dying wish that he might witness and assist its completion.

W. H. Monk describing Sir Henry's interest in the 'Psalter'.

Dear Sir Henry

You have broached a very difficult question. I think the book would appear better without bars at end of lines of words – but I believe them to be an absolute necessity. So *I vote for them.*

As to the expression marks – please do not introduce *too many* or all will be overlooked. Moreover genuine congregational singing is hopeless where a choir is alternatively rising on a high wave of sound and suddenly dipping into its trough.

I was obliged to use *Thou art gone up* in S. Paul's last Ascension Tide – so I dished up your tune as enclosed. It seemed to go well & to please.

I send one to Monk and to Dykes. I think it would be better to alter the harmonies (as I have) and so make it suitable to the words – than to kick it out because it is unsuitable.

May God's blessing be with Dr. Dykes in his important struggle for true Christian liberty. My wife sends kind regards.

John Stainer.

I like the type very much. You *must* mark *Unison* and *Harmony* where the effect is good.

Letter and postscipt from John Stainer to Sir Henry Baker simply dated Sept. 17

Dear Sir Henry,

I return the proofs with this.

We ought I think to come to some understanding as to the harmonies of the Amens.

As a rule I prefer the latter, if the tune is not in a higher key than D. In the proofs – they seemed to be used indiscriminately.

After sleeping over it – I am inclined to think that the suggestion I offered for the last line of my Tune to *My God I love Thee* – is *not* an improvement on the original which had better after all stand as printed.

I have not yet got my proof of 'The roseate hues' – but if it does not soon arrive I will write again for it.

Please do not introduce Sudeley into H.A.& M. – the inner voice-parts run about too much. I do not like the Tune, and would rather try and write you a new one if a C.M. is required.

I hope to be at my home from the 17th – 23rd inst.

Yours truly,

J. Stainer

My best remembrances to Miss Baker.

My dear Sir Henry,

At length I address you on the subject of the sum which I am to receive for my labour on the Appendix. It is not one of those subjects on which one likes writing and I delayed it, at first, in the hope you would yourself make the proposition: and latterly from a simple unwillingness to talk or write about it.

But it is time something was said about it, and moreover I now write because I wish to be favoured with your reply.

I have received £100 – I think the payment the Committee should make me, *now*, should be £150 more; making £250 in all, in present money, and then I think the final sum should be double that total. That is, that I should receive £500 in all. Considering the application I have had to make to the work – certainly equal to that bestowed on the original book and the long date back at which the application began, I do not think I am asking too much.

I can fairly say that the attention I have given has not been unattended with sacrifice of other things; as you yourself know: and indeed my receipts for this year have been so much affected by it (in postponement and refusal of private pupils) that I am now really in want of the second remittance now asked: and I trust it may be convenient to you to favour me with it in the course of a post or two.

I propose that the other half of the total sum be paid me whenever the Committee please to consider it due, and it is convenient to them to pay it.

Kindly do the best you can for me.

Ever yours very truly

W. H. Monk, the Music Editor of the Appendix to 'Hymns Ancient and Modern' writing to Sir Henry from Stoke Newington in January 1869

o sooner was the new decade begun that the restless roprietors were looking ahead again to a Revised Edition f their Book. In this connection, one of Henry Baker's tasks as to try and persuade authors to accept amendments to iosen texts which the Editorial Committee thought in their isdom would be desirable. In 1870, Caroline Noel, herself ı invalid, published an enlarged edition of *The Name of sus, and other Verses for the Sick and Lonely* containing e hymn *At the Name of Jesus*. With a tune EVELYNS ritten for it by Monk, the Proprietors wanted to print it ith changes. The reply dictated from her sickroom by aroline, clearly known to Henry in childhood years, shows e sensitivity which such correspondence required. Would *ernal* have been better than *mighty* at the end of verse 1? ut, though, for their request we should not be singing -day *But with awe and wonder.*

Co-proprietor Reverend George Cosby White, sometime Chairman of the Proprietors, was educated at Trinity College, Cambridge; he served in 1848 and 1849 as Curate of Wantage, Berkshire and from 1851 to 1853 as Provost of Cumbrae, Argyllshire, where he edited *Hymns and Introits* (1852). From 1853 to 1856, he was Francis Murray's curate at Chislehurst, Kent, and then appointed, first as Curate-in-Charge and then as Vicar of St. Barnabas' Church, Pimlico where he remained until 1876. He served as Warden of Beaufort Almshouses from 1877 until his retirement to Cleveland, where he lived to reach the age of 93. The Pimlico Church had been the centre of demonstrations against High Church practices in the days of his predecessor and these continued during White's ministry. In his own day the *ritual of the services* was *beyond the level of that* used in the mother church. Pimlico was then a poor-class area of London yet *on Easter Day, 1858, the throngs of communicants were so great that the church was quite incapable of containing them . . .* there were 483 communicants.

Cosby White used his share of the profits from *Hymns Ancient and Modern* to found St. Barnabas' Hostel, at Newland near Malvern, for aged priests of small means. Four houses were built in which the priests lived free of rent and rates.

The Communion Offertories at Monkland in 1872 were principally used as follows:

Coals for the Poor	**£9. 9. 9.**
Clothing Clubs	**£8. 3. 2.**
School Fund	**£21. 12. 5.**
St. Paul's Cathedral Fund	**£2. 0. 8.**
St. Martin's Home	**£3. 1. 7.**
Hereford Infirmary	**£4. 12. 5.**
Herefordshire Church Union	**£2. 5. 6.**
Society for Propagation of the Gospel	**£3. 10. 3.**

Dear "Harry" – for it seems impossible to call you by any other than that childish name, by which I have always remembered you – your kind and pleasant letter was a gt surprise & gratification – that anything I have written should be thought worthy of a place in that dear book, & should thus be used in Divine Service is to me a cause of gt thankfulness.

But you must forgive me if I say I don't like the alterations that have been made – was it necessary in the 1st verse last line to change, *mighty* to *eternal*?

In verse 4, the meaning is entirely altered – some one has changed it to

Bore it up resplendent
With redemption's light

What I intended to express was, that the Name wh for 33 years had been a household word amongst the lowly, the lost, & the suffering, *was borne up unchanged to the highest heavens*. If you don't like the words as they stand in my book, I shd not mind altering them thus:

Bore it up still wearing
Its soft (or mild) human light.

In the 5th verse I fear I cannot alter the second line – but the 3rd line could be changed to:

But with awe and wonder

I fear I cannot make any other alterations – my head is peculiarly weak just now – and there are some persons who wd greatly dislike that 'Brothers' should be altered in the last verse.

And now I fear I *must* end, for I am not equal to dictation, though there is a great deal I should like to say. Believe me, with affectionate remembrances,

Your old friend

C. M. Noel

This letter about 'At the Name of Jesus' was dictated by the writer from her sick bed

Dear Sir Henry Baker,
Thank you for your 2 notes, which have followed me here. I confess it never occurred to me that the Socy. was taking an undue advantage in the low terms it offers. I am a little puzzled, & do not quite see it now. But you may be right. I had never thought about it at all. As to the Socy. of S. Timothy, I wd. gladly try to persuade our Bishop, but he is, I fear, both narrow and obstinate. I shd. have greatly liked to have a branch myself, but its place is occupied to some extent by a little Guild of Holy Living (I must not tell this Bishop), & that, with a Guild of Church Workers, is almost as much as we can manage in a country village.
In great haste,
Ever yours very truly
Wm. Walsham How
I wish all Ember Weeks were kept as here.

Hymn writer William Walsham How writes to Sir Henry Baker from Lincoln in December 1875 about copyright fees charged by the Society for Promoting Christian Knowledge and about parish matters.

An important rival to *Hymns Ancient and Modern* wa[s] *Church Hymns* published by the Society for Promotin[g] Christian Knowledge. Arthur Sullivan was Music Editor o[f] *Church Hymns* (with William Walsham How and Joh[n] Ellerton co-editors of the words).

(v1)
At the Name of Jesus
Every knee shall bow,
Every tongue confess Him
King of glory now;
'Tis the Father's pleasure
We should call Him Lord,
Who from the beginning
Was the mighty Word.

(v2)
At His voice creation
Sprang at once to sight,
All the Angel faces,
All the hosts of light,
Thrones and Dominations
Stars upon their way,
All the heav'nly Orders,
In their great array.

(v5)
Name Him, brothers, name Him,
With love as strong as death,
But with awe and wonder,
And with bated breath;
He is God the Saviour,
He is Christ the Lord,
Ever to be worshipp'd,
Trusted, and adored.

(v7)
Brothers, this Lord Jesus
Shall return again,
With His Father's glory,
With His Angel train;
For all wreaths of empire
Meet upon His brow,
And our hearts confess Him
King of glory now.

Caroline Noel (1817 - 1877), who began writing hymns at the age of 17. She was the youngest daughter of Rev. the Hon. Gerard T. Noel, a Vicar of Romsey.

In correspondence with William Walsham How, Baker was extremely firm in protecting A & M copyrights. On the other hand he wanted Sullivan tunes for the enlarged Revised Edition of his book; and seems to have been successful in his negotiations. Sullivan had made sure he held on to many of

Dear Sir Henry
Your kind note has been forwarded to me here, where I am enjoying myself for a few days longer, before the Liverpool Festival. Can you not manage to run over and hear the *Light of the World*? **Thursday Oct. 1. Try.**
I am glad you like my tunes. Of course you are at liberty (as you know) to select any ten from our book, over which I have the control. *Resurrexit* **is by far my best tune musically speaking.**
Yours very sincerely
Arthur Sullivan writing from Sir Coutts Lindsay's in Fife. September 1874

his own copyrights. After the Revised Edition was published in 1875, the Proprietors refused permission for the use of certain copyright material to Rev. E. H. Bickersteth, who edited the principal other rival hymn book. For the Revised Edition Baker wrote the quietly strong text *My Father, for another night* to his own tune ST. TIMOTHY. He also wrote *O perfect life of love!* The text was written in Monk's house, where it was the subject of extended discussion one evening. The tune ABER was conceived by the composer in his sleep that night: by the time he met the author at breakfast, Monk had committed it to paper and was able to sing it to his guest. The tune is also used for another of Baker's texts *O Holy Ghost Thy people bless* which he wrote originally for a small A & M venture *Hymns for the London Mission* produced in 1874. It is also sung to CLAPTON and ALBANO.

My Father, for another night
Of quiet sleep and rest,
For all the joy of morning light,
Thy Holy Name be blest.

Now with the new-born day I give
Myself anew to Thee,
That as Thou willest I may live,
And what Thou willest be.

What'er I do, things great or small,
What'er I speak or frame,
Thy glory may I seek in all,
Do all in Jesus' Name.

My Father, for His sake, I pray,
Thy child accept and bless;
And lead me by Thy grace to-day
In paths of righteousness.

Henry Baker

O perfect life of love!
All, all in finish'd now;
All that He left His throne above
To do for us below.

No work is left undone
Of all the Father will'd;
His toils and sorrows, one by one,
The Scriptures have fulfill'd.

Henry Baker: Opening verses of Passiontide Hymn

O Holy Ghost, of sevenfold might,
All graces come from Thee;
Grant us to know and serve aright
One God in Persons Three.
Closing stanza of 'O Holy Ghost, Thy people bless'

The question of copyright was a thorny one. Many felt that hymns were not (and certainly should not be) protected by the Copyright Act of 1842. But they were! The question of who would eventually own the *Ancient and Modern* copyrights proved even thornier. The original Deed of Partnership meant that, if Proprietors were not replaced, the benefits would finally fall to the estate of the last surviving trustee. After Ward's resignation in 1863, the team of ten was maintained until the Rev. W. U. Richards died in 1873. Baker was not anxious for him to be replaced – perhaps so large a number of Proprietors was a bit bothersome to the man who did most of the work – and instead of making an appointment, various ideas were discussed in Committee. The deaths of Maberley and Harrison and Baker in 1878 reduced numbers to six, of whom Wilkins was of unsound mind. Whilst a new Deed was prepared in 1878, the Committee of Inspection refused to sign and it was possibly invalid on this account. Not long before, Lyall's Church of St. Dionis Backchurch in the City of London was closed as redundant after an extensive and intense dialogue between Lyall and his ecclesiastical superiors. He felt the Church Authorities were wrong to allow the removal of remains from the churchyard; and generally that he had been let down after serving the church for many years. Anyway, by the end of 1879, he left the Church of England and joined the Church of Rome. His co-Proprietors were flabbergasted, and considered that he should automatically resign. He felt differently, probably needing the income as a Proprietor, which was his legal right. There was much acrimonious correspondence, with Lyall finally agreeing not to attend meetings if he received notices of them and his share of distributions. He refused however to admit any new Proprietors.

Henry Baker's life-work was in a sense in the melting-pot. If Lyall were the last survivor, how could a Roman Catholic become sole owner of the book put together by clergy of the Church of England especially for that church? Indeed, the collapse of the market was a real possibility. The parties were involved in legal action, but the matter was finally settled out of court. Lyall was persuaded that the income would collapse even before he became a possible last survivor if his Roman Catholic status were generally known: and in the end he

Co-proprietor Reverend William Hearle Lyall was educated at St. Mary Hall, Oxford and Wells Theological College: after serving as Curate of Christ Church, St. Pancras, London, he served as Rector of St. Dionis-Backchurch in the City of London from 1853 until its closure in 1877; in 1876-77 he was President of Sion College, the society of Anglican clergymen possessing a great theological library. Like Christopher Harrison, Lyall was a member of the Society of the Holy Cross. After losing a bitter battle with the Authorities over the proposed closure of St. Dionis-Backchurch, he left the Anglican priesthood and gave allegiance to the Roman Catholic Church.

. . . And what is soon to follow on this spot, where we now meet for the last time; whose congregations have worshipped for at least six hundred years, whose successive generations have been interred and gone to their *long home*? All that is sacred is to be swept away, and the place given up to secular purposes. *(Final Sermon)*

. . . I cannot think that the end justifies the means, or outweighs the violence to sacred principles and hallowed associations, especially, too, when the wealth and ample resources of the Church of England and her members are taken into account. (Campaign statement.) It was proposed to use the funds from sale of the site to build churches in new areas of population.

agreed to do what was needed of him provided he continued to receive the same financial rewards for the rest of his lifetime. Henry Baker's great work was saved. A new Trust Deed of 1895 ensured that for the long term newly elected Proprietors would essentially be Trustees (and incidentally, could not serve unless they continued to be members of the Church of England). So Henry Baker's work has continued to serve this present age, as his book has been constantly revised to meet the needs of succeeding generations.

In 1876 offerings were received in Monkland Church for the Dykes Memorial Fund (£3. 17. 11) and the School Fund (£22. 6. 11½)

Sir Henry's last hymn was written during a holiday in Ireland and is dated Killarney, September 1876. *Redeem'd, restored, forgiven* was first printed in the *Monkland Parish Magazine* in November, and in his final anthology *Hymns for Mission Churches* published in 1877. Tunes are PASSION CHORALE (O HAUPT VOLL BLUT UND WUNDER; HERZLICH THUT MICH VERLANGEN) and TOURS. For Reverend Sir Henry Baker, Bart. his earthly travail was almost over. On 3rd January 1877 he was taken ill, and although he recovered enough to write a pastoral letter on 30th January he died at Monkland on 12th February. The pall was borne by the majority of his *Ancient and Modern* colleagues with William Monk leading the music in the churchyard of the Church Baker had himself seen restored. His hymn *The King of Love my Shepherd is* was sung: it includes the verse which he spoke as the last audible words of his mortal life:

Perverse and foolish oft I strayed,
But yet in love He sought me,
And on his shoulder gently laid,
And home, rejoicing, brought me.

A stained glass window erected in the South Wall of the Chancel Chapel depicts scenes from the hymn.

. . . it was to his unflagging zeal, his perhaps unrivalled knowledge of hymns, his singularly retentive memory, his hearty love of the work itself, that the marvellous success of H.A. & M. may be said to be mainly due . . .
. . . it can only be with a feeling of unmixed satisfaction that we can look back through the course of so many years, and acknowledge with the deepest thankfulness that, if there were at times passing differences of opinion, there always subsisted an inward cordiality and unbroken friendship between himself and his colleagues that perhaps has never been exceeded.

Extracts from the Minutes of the Proprietors of 'Hymns Ancient & Modern', April 12, 1877

The beautifully restored church with its excellent choir and crowded services, the efficient school, and indeed every department of the parochial machinery bear witness to the successful zeal of the late vicar and his devotion to the duties of his sacred calling. His memory will live long in the grateful recollection of the people of Monkland whom he loved so well and for whose good he laboured so heartily up to the beginning of his short illness.

A tribute to the life and work of Sir Henry Baker, Bart., printed in the 'Hereford Journal' of 17th February, 1877

. . . not only among those musicians who had the good fortune to hold intercourse with him will his hearty warmth and genuine affection be ever missed, but also by the comfortless widow, the aged, and the orphans in that small parish which was his home. With every face he was familiar, sympathised with every suffering heart, joyed in every smile; and no dissentients did he find to the beautiful forms of divine worship carried out in his church after his own liking, because all who knew him were drawn to him by his simplicity and purity of character, and loved him with a love which casteth out fear.

From a report of Sir Henry Baker's death in the 'Hereford Journal'

Redeem'd, restored, forgiven,
Through Jesus' precious Blood,
Heirs of His home in heaven,
O praise our pardoning God!
Praise Him in tuneful measures,
Who gave His Son to die;
Praise Him Whose sevenfold treasures
Enrich and sanctify!

Once on the dreary mountain
We wander'd far and wide,
Far from the cleansing fountain,
Far from the pierced side;
But Jesus sought and found us,
And wash'd our guilt away;
With cords of love He bound us
To be His own for aye.

Dear Master, Thine the glory
Of each recover'd soul;
Ah! who can tell the story
Of love that made us whole?
Not ours, not ours the merit;
Be Thine alone the praise,
And ours a thankful spirit
To serve Thee all our days.

Now keep us, Holy Saviour,
In Thy true love and fear;
And grant us of Thy favour
The grace to persevere;
Till, in Thy new creation,
Earth's time-long travail o'er,
We find our full salvation,
And praise Thee evermore.

Henry Baker

Lord, from whom beauty, truth and goodness spring,
World beyond shining world proclaims Thee King;
Earth in her life and loveliness displays
Thy Spirit's presence and declares Thy ways.
With all creation we our worship bring.

For all the splendour changing seasons hold,
Snowfields of winter, autumn's harvest gold,
Spring's life resurgent, summer's feast of flowers,
Morn's waking glory, calm of evening hours,
Our thanks, O Father, ever must be told.

Dwelling in this fair town upon a hill,
Spire soaring high beside an ageing mill,
Homes bowed with years along a curving street,
Down to the Guildhall where old highways meet;
Where faithful hearts adored, we worship still.

Thine are the riches of our heritage,
Music and dance, carved stone and author's page.
Thine too the gifts of home and faithful friends,
Joys of true comradeship Thy goodness sends,
While in Thy service, Father, we engage.

Chiefly we praise Thee for immortal joy
Christ gives and nevermore can death destroy:
Joy of eternal fellowship with Thee,
This life's sweet foretaste of the life to be,
When praise will ever all our powers employ.

Albert Bayly: 'A Hymn for Thaxted'

Albert Bayly

Albert Frederick Bayly loved to scale the heights and to plumb the depths. Out walking, the crest of a hill was no final destination if another summit was there brought into view: for all his concerns about God's kingdom on earth and the everyday needs of humanity, his faith, as he walked through life, looked further to the country beyond death. In his study and in his conversation, he sought immersion in the depths of the eternal mysteries: here too was a radical who saw the marvels of science rooted in the love of God.

The year of his birth coincided with the first trans-Atlantic radio message when Guglielmo Marconi sent signals successfully from Poldhu in Cornwall to St. John's, Newfoundland. It was the year, too, when adrenalin was first manufactured. Bayly, sometimes called the first hymnwriter of the moderns, was in sympathy with a century of scientific and medical discovery. In 1901, also, Lord Kitchener used a scorched earth policy against the Boers. Such waste, and all the evils of war, were anathema to Bayly, who saw pacifism as a natural outworking of Christian commitment.

Albert Bayly

Born on England's south coast at Bexhill-on-Sea on 6th September, 1901, he was the eldest child of his father's second marriage. He had two half-brothers and a younger brother and sister. His family were active in the local Congregational Church into whose membership he was received at the age of 13. He attended Hastings Grammar School, leaving early to train in Portsmouth as a shipwright. His time in the Royal Dockyard School and his work in the Yard gave him wide practical experience. His family moved in the meantime to Waterlooville, and there being no nearby Congregational Church, they attended worship with the local Baptists.

This serious young man was fascinated by the stars and set off to the local public library to discover a book that would introduce him to astronomy. The volume suggested by the assistant was certainly about stars – film stars. Albert Bayly's library was in due course to feature many books on the sky's heavenly bodies: his hymns and poems are full of references inspired by the wonders of the cosmos. He writes of *teeming particles turned to ordered tasks, forming the galaxies with myriad suns*, of *rays that span in light-years the distances of space*, of God, *whose majesty outshines Heaven's dome of stellar light*. His hymn, *Lord of the boundless curves of space*, written in January 1949, inspired Derek Williams' tune SAN ROCCO. The original text was sparked off by a talk on radio by J. Isaacs on *Poetry and Science*. In turn, Bayly's first three verses brought three new ones as a replacement ending from the pen of Brian Wren. Now Bayly's opening and Wren's finishing form the usually sung version, with Albert readily agreeing the new ending improved his text. Besides SAN ROCCO, tunes used for the hymn include LONDON NEW, SALZBURG (Haydn) and CAITHNESS.

BAYLY'S ORIGINAL VERSION

Lord of the boundless curves of space,
And time's deep mystery;
To Thy creative might we trace
The fount of energy.

Thy mind conceived the galaxy,
The atom's secret planned;
And every age of history
Thy purpose, Lord, has spanned.

Thy Spirit gave the living cell
Its hidden, vital force:
The instincts which all life impel
Derive from Thee their source.

Thine is the image stamped on man,
Though marred by his own sin;
And Thine the liberating plan,
Again his soul to win.

Science explores Thy reason's ways,
But faith draws near Thy heart;
And in the face of Christ we gaze
Upon the Love Thou art.

Christ is Thy wisdom's perfect word,
Thy mercy's crowning deed:
In Him the sons of earth have heard
Thy love's deep passion plead.

Give us to know Thy truth; but more,
The strength to do Thy will;
Until the Love our souls adore
Shall all our being fill.

THE COMPOSITE VERSION

Lord of the boundless curves of space
And time's deep mystery,
To your creative might we trace
All nature's energy.

Your mind conceived the galaxy
Each atom's secret planned,
And every age of history
Your purpose, Lord, has spanned.

Your spirit gave the living cell
Its hidden, vital force:
The instincts which all life impel
Derive from you, their source.

You gave the growing consciousness
That flowered at last in man,
With all his longing to progress,
Discover, shape and plan.

In Christ the living power of grace
To liberate and lead
Lights up the future of our race
With mercy's crowning deed.

Lead us, whom love has made and sought,
To find, when planets fall,
That Omega of life and thought
Where Christ is all in all.

The serious young man was not drawn just to astronomy but to ministry. He was also now convinced that pacifism wa the right road for a Christian and this was clearly at odd with his work in the Royal Dockyard. Albert Bayly sought out the Revd. Henry Parnaby, the minister of Buckland Congregational Church, Portsmouth, to guide him about offering as a candidate and was advised to take a correspondence course towards an external Bachelor of Arts degree. As he pushed his cycle up the hills to and from work, he battled with Latin vocabulary. In 1924, he received his degree.

On the 2nd August 1914, the day before the outbreak of World War I, a significant meeting at Constance set up the *World Alliance for Promoting International Friendship through the Churches.* The initiative was followed up at The Hague in 1919 with a proposal which led to the Ecumenical Conference of Protestant Churches in Stockholm concerned with Life and Work in August 1925. When Bayly took up residence at Mansfield College, Oxford that autumn, ecumenism was a live topic. The Conference had found much common ground, but there were substantial differences of approach on birth control, temperance and war.

In conclusion, I should like to say one word about our own country. Nothing is more essential among us than, to use the words of Dr. Söderblom, *a common voice for the Christian consciousness, and a common organ for Christian action.* **Corporate union is scarcely within sight. Even federation of the Churches of Christ in our midst is, perhaps, not yet practical. But surely co-operation in connection with problems of Life and Work should be possible in England and in Wales. And it should not be too much to expect the co-operation to embody itself in a visible symbol – be it a permanent committee or anything else – which would inspire and guide concerted action among all the Churches within the fellowship.**

From an article by E. O. Davies in the December 1925 issue of the 'Mansfield College Magazine' reporting on the Stockholm Conference on Life and Work.

Six of the fifteen freshmen at College that year came from overseas: three from the United States, one from South Africa and two from Canada. Significantly, the year marked the establishment in Canada of the United Reformed Church bringing together Methodists, Congregationalists and some Presbyterians. A pre-terminal retreat in Bayly's first term included a paper on *Child Psyscholgy and Modern Sunday School Methods.* In outdoor vein, Bayly and two others from the House were associated with Cat's Boat Club.

In the Hilary Term, a Retreat looked at Race problems, while a Quiet Day looked at a paper presented by J. C. Bennett on *Doubt.* We can share a college experience of the young theological students of Albert's day as we wrestle with a few extracts.

> There is only one doubt that really matters very much, and that is doubt about the existence of the Christian God. If a man has a God in whom he can trust, none of the other possible doubts need distress him.

. . . . .

> It seems to me to be fundamental to any consideration of doubt to make a distinction to start with between two kinds. I mean that this one doubt about the Christian God has two causes, or two sets of causes. The first kind is caused by some weakness within the individual doubter, which blinds his soul to spiritual

. . . That Confucianism has weaknesses, exaggerations and positive errors, is obvious to all, and arising out of these has come the present revolt of Young China against it and all its works. On the other hand we must not be blind to its plain achievements and ultimate truth, specially as purified and enriched in the light of the Christian revelation.

From an article in the 'Mansfield College Magazine' of December 1926 on 'Spiritual Authority – A Chinese Approach to the Question' by E. R. Hughes

things. The second kind springs from something in the world, which the mind of the doubter cannot reconcile with the hypothesis of faith. Now both of these two kinds of doubt are usually present at once, in varying degrees. And when we come to ask the question, what should be our attitude towards our doubts, the first answer is, that it makes a great deal of difference which kind of doubt predominates. It is difficult to find this out in any particular case. These two causes of doubt interact and reinforce each other, so that they are not easily separated. To add to the difficulty, when a state of doubt is due to the first cause, it always claims to be due to the second. That is to say, when a man's doubt is caused by a disorder in his digestive system, he sincerely thinks that it is caused by a disorder in the nature of the universe.

. . . . .

An outstanding modern exponent of this point of view is Frank Buckman. There is probably no religious leader among undergraduates in at least the eastern part of America who has more influence, for good or evil, than he. I know of no man who is more enthusiastically admired, or more violently hated, as a result of the thoroughness with which he holds to this idea, that doubt is always caused by sin.

Extracts from J. C. Bennett's paper on 'Religious Doubt' printed in the June 1926 issue of the 'Mansfield College Magazine'

. . . We remember her wise and moving talks in Chapel, revealing the depth of her personal faith and her sympathetic understanding of human need. Above all, we admired her dauntless courage and triumphant cheerfulness in face of physical handicaps which would have overwhelmed many people. Not a little of her college work was done from a sick bed. This ill-health pursued her from the age of fifteen to the end of her life. . . She once said: *There is no theoretical answer to the problem of suffering, but there is a practical answer.*

Albert Bayly writing in 'Mansfield College Magazine' about the life of his fellow-student, Dorothy Wilson.

Other reports in the *Mansfield College Magazine* tell of a retreat during Bayly's time when the subject on Friday evening *The Construction of the World through Faith* led the students into deep waters. Another retreat considered *Psychology and the Subjective View of Religion.* A conference took a provocative look at Foreign Missions, whilst a lecture about China on *Spiritual Authority* by E. R. Hughes is just one example of studies with a world view. The excellent Christmas production in 1926 of the Burnand and Sullivan operetta *Cox and Box* provided light relief!

One of Bayly's contemporaries at College was Dorothy Wilson, the first woman to be a full member and to gain the Oxford Diploma of Theology. She became one of the first women ministers of the Congregational Church, but her many achievements were against the background of persistent ill-health. Bayly was to value return visits to Mansfield College over many years.

Back in 1901, a small group of Congregationalists, including some from nearby Newcastle, met at a house in Cullercoats with a view to establishing a Church in Whitley Bay. The first services were held in Eden Cafe on 16th June 1901. The cause prospered. An iron hall, known as *The Tin Tabernacle* was opened in 1902 and the Park Avenue Church built in 1907.

Whitley Bay Co-operative Society Limited, c. 1920

Mr. R. S. Darling, a young deacon of this church, saw a need for services in the growing township of Monkseaton which adjoined the Northumbrian holiday resort. These began at Fairway Hall, a building attached to Rose Cottage for the purpose by Mr. H. B. Saint, in February 1927, being conducted mostly by laymen and occasionally by the minister of the Park Avenue Church. The hall was also opened on Monday evenings as a Reading and Recreation Room. The need for further help in getting the cause off the ground led to the Principal of Mansfield College suggesting a student for this task. So Albert Bayly came to Monkseaton in July 1928. When he arrived attendance at the Sunday service was five, but by the time of his ordination in the following March the number had risen to twenty. A Sunday School had been started and there were fourteen on the Cradle Roll; twenty-three in the Beginners and Primary Department; another twenty-three Juniors and six Intermediate scholars, making a total of sixty-six.

That this Church Meeting is desirous of furthering the Kingdom of God in the neighbourhood of Old Monkseaton and now resolves to plant a Congregational Church in that district.

Minutes of Church Meeting of Park Avenue Congregational Church of 29th June 1927

It is appropriate for a saintly man that, on the occasion of his ordination, he was welcomed on behalf of the congregation by Mr. H. B. Saint. In his charge, the Rev. Dr. W. B. Selbie, the Principal of Mansfield College, reminded Albert Bayly that, as a Congregational minister, he was *called to be a teacher of God's people, unfettered by the tradition of the past, to proclaim faith in Christ . . . not only salvation for the future . . . also for men's present needs to help them to do their work . . .* He was to *be a pastor, a keeper of souls.* The new minister, speaking about his experiences in college, referred to Dr. Selbie as *more than a teacher*: he was pastor, friend and father.

'Have the former Free Churches lost their freedom through union with what was the Anglican Church?' I asked.
'On the contrary' was the answer, 'they have found a greater freedom through sharing a wider and richer heritage of Church life. Moreover, the former Anglican Church, freed from some of its formal ties with the State, has shared this experience of greater freedom, and is playing its full part in striving to make the Christian Faith, in fact and not merely in name the Faith of the nation.'
From Albert Bayly's dream of the English Church in the 21st century Written in 1965

Albert Bayly's appointment was as assistant minister for the Park Avenue Church where the ordination service took place. The event began with tea and an organ recital. The same issue of *The Whitley Seaside Chronicle and Visitors' Gazette* which reported the service featured as the largest advertisement on its front page this declaration:

The Youths are bright because the Buttons are bright
and the Buttons are bright because they are brightened
with CWS Liquid Metal Polish.
Sold by Co-operative Societies everywhere.

CWS stands for Co-operative Wholesale Society. These were days when Co-op trading was in its ascendancy, particularly in the North.

The same week the Urban District Council announced a reduction in the rates, whilst at local cinemas, alongside Ramon Navarro and Norma Shearer in the film of *The Student Prince*, Bebe Daniels in *The Fifty-Fifty Girl*, James Murray in *In Old Kentucky*, and Lilian Harvey in *Matrimonial Holiday*, one house featured *At the End of the World*, a gripping drama in which Brigette teaches a lesson in the folly of war. The Congregational Church Dramatic Society presented Barrie's *What every Woman knows* and the Minister, the Rev. E. Neville Martin, who had participated in the Ordination service, made a very good first appearance as an amateur actor.

Whilst most of Albert Bayly's work was to be at Monkseaton, he preached on Whit Sunday at the main Church and again the following week when the theme of Medical Missions was one especially close to his heart. In June, he began Monthly Young People's Services at Fairway Hall with themes such as *Christianity and Leisure, Ambition, Youth and World Peace* and a three-part series on *The Romance of a Book in the Making.* Alongside the cinema, the lantern lecture persisted, with the Rev. Iona Wills extending local horizons with a talk on *The Riviera.* At Harvest, worship was enriched by a cellist. Albert Bayly was always ready to encourage musicians to play in church. By April 1929, there were ten Cubs and seventeen members of the Girls' Friendly Society.

On 24th October 1929, the Wall Street collapse triggered an economic crisis that was to lead to the harsh days of depression which continued through the thirties.

In the autumn of 1930, young people's services at Fairway Hall were advertised on *India* reflecting a life-long concern for the world church and on *Patriotism* on which subject his pacifist views had a bearing.

A meeting in December at which the Rev. C. Leslie Atkins, Minister of St. James' Congregational Church, Newcastle, was invited to give an address *The Mysterious Universe* confirmed Bayly's enthusiasm for stretching minds on the nature of the cosmos.

Monkseaton was growing fast and at the beginning of 1931 a press correspondent was campaigning for a public clock, a hardware business, a modern post office, a jeweller's shop and a public lecture hall. In April 1931, a Women's Guild was launched at the church. Heroes celebrated in Youth Services included Bayly's greatest hero of all, Albert Schweitzer of the African forest, the French theologian and missionary surgeon who, save for brief trips to Europe to play Bach on the organ to raise funds, gave his life for the peoples of Gabon where he founded a hospital at Lambarene in 1913. Mohandas Gandhi, the great advocate of non-violence, and Arthur Peill, beloved physician of Tsangchow, were also celebrated in worship in 1931. Locally he welcomed Cullercoats Fishermen's Mission Male Voice Choir to Fairway Hall. By the end of the year, attendance at services had reached 65; and there were 111 on the Sunday School Roll, of whom 65 on average attended each week.

Schweitzer, thou royal hearted friend of man;
Great in the realms of music, healing, thought;
Clothing in deeds the wisdom thou has taught,
Shaping thy life to youth's heroic plan;
Prophetic soul, whose thinking far outran
The spirit of thine age, and questing, caught
Vision of truth humanity has sought
Since mind's immortal Odyssey began:

Honour we gladly pay thee: but thy praise
Rings strangely in this world of angry strife:
A nobler song of tribute we shall raise
When, from this age with bitter passions rife,
We turn to wiser deeds and kindlier ways,
And keep thy precept, reverence for life.

Albert Bayly: 'To Albert Schweitzer' inspired by reading George Seaver's book 'Albert Schweitzer, the man and his mind'. February 1948

When the Rev. S. A. Evans came in May 1932 as the new minister at Park Avenue Church, Albert Bayly said he had come among a people who would reward him by whole-hearted co-operation and faithful loyalty. Although the people of Tyneside perhaps did not say very much at first, he would find there were no people in Britain with warmer hearts or a greater capacity for loyal and devoted service. In later life, as his ministry took him around England, Albert Bayly was to think of the North-East rather than his native South coast as home. That year, the World Church focus was on Papua; a lantern lecture featured life and

Albert Schweitzer

religion in India and he chaired a meeting at Whitley Bay for a visit by Mr. Shoran S. Singha, an Indian who was the grandson of one of the first converts of the pioneer missionary, Dr. Alexander Duff.

Thine is the quest of man's aspiring mind,
Climbing through mystery to God. The pain
And peace of self renounced to find
Fulfilment in the infinite again.
Thine is the pride of empire: Moslem, Greek
And Briton battled for thy throne.
Thine is the greater glory of the meek,
In humble service to the needy shown.
Thine, India, with the shame of ancient wrong,
Are golden deeds of justice, mercy, grace:
All loveliness of colour, form and song:
High aspirations of a noble race.
Yet shall thy peoples' glory brightest be,
When Christ to all brings Life – full, joyous, free.

From a pageant 'God's Building' written by Albert Bayly commemorating the story of the Church in South India published in 1949 by the London Missionary Society

After an evening service in February 1933, there was a public meeting of the League of Youth under the title *Youth and the League of Nations* with the Fairway Hall Choir providing musical items. Earlier, after another evening service, the congregation unanimously passed a resolution on Disarmament and sent it to the local branch of the League of Nations Union.

On 26th April, 1933 Albert was married to Marjorie Shilston, the eldest daughter of Mr. and Mrs. R. D. Shilston of West Hartlepool. The ceremony was conducted in London by Rev. Dr. A. Herbert Gray at Kingsley Hall, Bow. The congregations of Park Avenue and Fairway Hall presented the couple with a Hall Wardrobe for the Manse.

In 1934, Albert Bayly headed two campaigns. He organized a house-to-house canvass to gain recruits for the League of Nations Union, and some 250 new members were enrolled. Some of them took part in September in a Pageant and Witness for Peace in Newcastle-upon-Tyne. In October, the Monkseaton Church staged a Rally Week. This involved visiting everybody who had the slightest connection with the Fairway Hall. It began with the dedication of Sunday School teachers and featured the Guides, Brownies, Scouts and Cubs with special events every night of the week. The same year gifts were sent to the Polish Flood Relief Fund and gifts of toys for children of Nazi families in Germany were made with the help of the Society of Friends in Berlin. In lighter vein, February 1935 saw a presentation of the pantomime, *Cinderella.* Not many of the Park Avenue congregation took much interest in the sapling they had planted at Fairway Hall. Perhaps this was especially so because the financial burden of providing most of Albert Bayly's stipend came from one man's pocket.

As war clouds began to gather over Europe, Bayly pursued his pastoral ministry, which included working with the Town's Special Aid Committee for the relief of poverty. He wrote scripts for children's events. All his life he believed that drama had a powerful contribution to make in the communication of Christian truths. He continued to be especially active in the work of the League of Nations Union which, in 1937, established a kiosk on the Promenade. Presenting his report that year, as Secretary of the Local Branch, he referred to *a series of events in world affairs which had undoubtedly given a severe setback to the work of the League and to all efforts for international peace. Present conditions, however, made the task of the League of Nations more, rather than less, necessary for it was vital, when so many people were reverting to old illusions of security in arms, to insist on the principles on which real peace and security must be built.* That year, in May, Bayly was organiser of a Peace Week, which included an exhibition with vivid posters depicting the ghastliness of bombing and aerial warfare. There were, in contrast, photographs showing the use of aircraft aiding explorations in the Antarctic, destroying pests by spraying and bringing infertile land into use. He wrote to the Press hoping that many would join in a Saturday demonstration but, despite excellent weather, around fifty people formed the total company.

An open-air peace demonstration has been arranged . . . Will members of local churches, guilds and other organisations, as well as the general public, please assemble at this time and place with any flags, banners and peace posters available . . . Let us make this procession and meeting a clear demonstration that Whitley Bay is resolved to work for peace.

Albert Bayly writing in May 1937 to the 'Whitley Seaside Chronicle and Visitors' Gazette' on behalf of the Whitley Bay Peace Week Committee

Fairway Hall continued as a Worship Centre at Monkseaton, at first with continued success. Wartime conditions brought real difficulties: in April 1941 the premises were damaged by enemy action. By 1946, only a handful of members remained and the site was sold around the end of that year. In 1975 the Park Avenue Church at Whitley Bay joined with St. Cuthbert's Church to form Trinity United Reformed Church, using the former Presbyterian premises.

On the 25th March, 1938 the members of Dacre Street Congregational Church, Morpeth, who had requested their previous pastor to resign in June 1937, invited Albert Bayly to be their new minister. Morpeth is an ancient town lying in the valley of the River Wansbeck. For centuries, tanning was the principal industry. It had seen fighting in the days of the Civil War: but generally had enjoyed a tranquil existence, being a very pleasant place to live and much less industrialised than Ashington, Bedlington and the Tyneside towns. The first Independent Chapel was hidden in a little-known quarter of Morpeth in King Street. Close by was a brewery, a tallow candle factory and two slaughter-houses. Divine Service was accompanied by the braying of donkeys and the barking of dogs. In 1898, a church was built in Dacre Street to seat 300 persons with a hall erected for a similar number.

At the Induction Service in Dacre Street on the 6th May, during those crisis-torn days in international affairs, he preached from the beginning of Isaiah 61, with its promise of the New Jerusalem, of good news, healing, comfort and liberty. *I remember the joy that swept over the country in November, 1918, when it became known that the Armistice was signed, and that the nightmare of war was over. What news would be more joyful over the world than that the danger of another world war had finally been averted.* A

That admirable handful of earnest and serious people who erected this Chapel built it, with commendable Christian humility, in what was regarded in 1829 as the worst district in the town. There the Independent Chapel stood, an island, as it were, in a sea of squalor and misery. Nevertheless this 'ugly Dissenting Chapel' was to many members, good and true, a veritable *Patmos* where the old men dreamed dreams and the young men saw visions.

From 'One hundred years of Congregationalism in Morpeth' published on the occasion of the centenary of the Dacre Street Congregational Church.

representative of the Fairway Hall congregation reported perceptively that *he had never heard Mr. Bayly preach but he found benefit from his sermons.* Marjorie Bayly was described as *an asset to any church.* When Neville Chamberlain brought back from his meeting with Adolf Hitler the false promise of *peace in our time*, many in Albert Bayly's congregation joined him in sending congratulations to the Prime Minister.

At the Harvest Festival, Bayly proclaimed that *religion can never become a vital thing till God's ploughshare has been driven deep among the hard-caked clods of our laziness, pride and insincerity. Till the soil of our life has been turned over so that the seeds of goodness can be planted in the heart of it and the rain of God's grace percolates to the depths . . .* For Europe, God's grace did not percolate into sufficient hearts to prevent, before the next Harvest Festival, the beginning of World War II.

Bridge Street, Morpeth, c. 1930

In November Albert Bayly was asked by the Deacons to make known from the pulpit the need of the brethren in the Czech church and to invite any friends who could to send him gifts to forward to the Church there.

The confrontation of good with evil came again in Albert's Advent sermon, reminding his hearers that the message of Christmas was for all nations and all ages, but that being a follower of the Crucified involved Stephen in martyrdom long ago – and this was the experience of African and Chinese Christians in 1938. Albert Bayly began the next year with a campaign for each Church Member to 'adopt' a child who came to church but the response was much less enthusiastic than the pastor hoped.

In 1939, the pacifist Albert Bayly, like so many with or without pacifist views, was horrified by the outbreak of war in September. Albert joined the Red Cross and served in a local First Aid Unit. From the very beginning Albert's home on Sunday evenings was always open to serving members of the Armed Forces in Morpeth. He maintained contact with some of these new found friends for the rest of his life.

Some of the church's noblest saints belonged to what was called the dark ages, when Europe was in a state of anarchy and wretchedness.

Words from a November 1940 sermon by Albert Bayly as reported in the 'Morpeth Herald and Reporter'

When preaching, Albert Bayly often told the life stories of his own heroes and in January 1940, emphasising the role of the pastor, recalled the work of Methodist minister, F. W. Chudleigh, who gave so many years of his life to serving alongside the people of London's East End.

Bayly offered the use of his home for a public meeting on *Pacifism and World Order*, called by the Morpeth Peace Group, a body affiliated to the Fellowship of Reconciliation. Some deacons objected to the use of Church property in this way, particularly in war-time, and Bayly explained that it was a *thorning* subject. The Group met the deacons' objection in the spirit of the peace movement, using a friend's house, feeling no resentment for the denial of the use of the Manse.

Despite the seriousness of war-time in Morpeth, there were many lighter moments: a poacher pointed out to the local court that 5,500 cabbages had recently been eaten by rabbits and hinted that his own activities might be in the service of feeding the nation.

Life demands one-tenth ability, and nine-tenths stickability.

From a United Service sermon on 29th December 1940 reported in the 'Morpeth Herald and Reporter'

As the Battle of Britain began, Bayly preached: *Christians do realise they belong together, even if they belonged to countries at war with each other. Even now, a World Council of Churches is being formed ready for the tasks which will await the Church when the war ends.* He reminded his congregation of a more immediate need: to open their homes to relatives of wounded men sent back from battle. In 1941, when news of the arrival of a few oranges strictly for children only hit the headlines and going on the beach meant a court appearance for disobeying a Government order, Albert Bayly quietly, yet energetically, continued his ministry. In February he preached on Red Cross Sunday.

If God can speak to us through others, He may also speak through us to them.

From a sermon of 25th May 1941 reported in the 'Morpeth Herald and Reporter'

There is God's ***ambulance*** activity; – agencies, many inspired by the Christian Church, dealing with immediate needs. The Red Cross and St. John ambulance were built up largely through Christian initiative and influence. Our Army medical services owe their quality to that noble Christian, Florence Nightingale. Modern ***first-aid*** to the poor, the out-of-work, the aged and blind, owes much to Christian influence. And God is still at work to-day in the midst of war wherever people are alleviating suffering, helping the homeless, the hungry, the prisoner of war and the interned alike. This is God's work, and in so far as we have the privilege of sharing it we can know that we are co-operating with God. But this is not enough to deal with the evils of to-day . . .

Jesus not only dealt with immediate needs . . . He taught men to love and forgive their enemies, – the real answer to war, for it deals with wrong in a way which cuts out the evil consequences of retaliation and revenge. Instead of destroying the wrong-doer, it makes him want to go good.

From a sermon for Red Cross Sunday in February 1941 as reported in the 'Morpeth Herald and Reporter'

When we look at Jesus in the wholeness of His personality in its historic setting, He stands unique. Here is the Teacher Who lived His highest precepts out to their last consequences on the Cross . . .

From a report in the 'Morpeth Herald and Reporter' of Albert Bayly's sermon on 15th February 1942 based on John 1, 1-14

Albert Bayly served on the Board of the London Missionary Society. In March an exhibition at the Manse about missionary work in the South Seas was punctuated by performances by the Pilots telling the story of the service of the ship *John Williams V* in the Islands. The Pilots, a junior missionary association of children supporting the London Missionary Society, was close to Albert's heart. Another constant concern for Bayly, was that Christian fellowship should degenerate into a social club and that keen young converts might find Church life drab and conventional. He was not afraid to speak his mind on this at a service in May. He spoke his mind again at Harvest Festival in 1942 when food rationing in Britain was at its height. During 1942, two deacons felt Albert Bayly should be replaced perhaps because of his continued pacifist views. But in general he received a vote of confidence save from these two complainants.

. . . This meeting having taken into consideration all matters relative to the Pastorate and the work of the Church in Dacre Street herewith records its confidence in Mr. Bayly and agrees to send him from this meeting a message of loyalty and affection from his people. Trusting in God, that we in full support with him, will endeavour to make Dacre Street a bright and shining light to the community.

Dacre Street vote of confidence in Albert Bayly 1942

Do we now show a lack of Christian imagination and sympathy in rejoicing over the satisfaction of our own wants in the world so full of suffering, while so many even of our fellow-Christians are starving. If there were nothing in our harvest thanksgiving but thoughtless satisfaction that God at least has provided for us there would be force in this question. But does not Christian thanksgiving demand something much deeper and finer, an acknowledgement of our dependence on God and his claims upon us which requires at least that we offer ourselves gratefully and humbly to further His purposes? If we want to thank God properly to-day we must not try to shut out unpleasant facts but look at the world with all its suffering and wrong and where we see God's Will being denied offer ourselves to him in any way we can to be used that His will may be done.

One fact which seems to deny God's will is that while our real needs are supplied there are multitudes whose needs are not. This year some of the richest lands of Russia have been devastated by war. Do we realise what that will mean to great populations in eastern Europe? Or take the situation in the occupied countries. A public man who recently escaped from Belgium says: 'We are approaching an acute crisis. Rations even if obtainable, provide only half the food needed, and they are not always obtainable. Over two million people have been without potatoes since December. Children suffer most. Few of the aged are likely to survive the war.' He says the main cause is the invasion and blockade. Before the war Belgium imported over two-thirds of her food. This has stopped. The Germans have seized stocks of raw materials and foodstuffs. The butter shortage is due to their purchases. But this should not be exaggerated. We must take into account the import from Germany and other continental countries of 400,000 tons of wheat in 1940-41 and a certain quantity of seed potatoes. The root of the trouble remains the loss of foreign trade.

The safe arrival and distribution of food imported by the Belgian Government from Portugal shows that something can be done from outside through the Red Cross without German interference, but is only a drop in the ocean. The crying need is for vitamins and concentrate food . . . A recent report says that unless suitable food is sent before winter the children of Greece are doomed.

We cannot have the right kind of harvest thanksgiving unless we take such facts into account. For God is not our God alone. He is the God of the whole human family. It is not His will that some should be satisfied while others want. Do we think He wants us to thank Him because we have more than others? Will He not say, let me use your help to put right a state of affairs so contrary to my will. Some believe that nothing can be done for the occupied countries till they are freed from the enemy. Others believe something can and ought to be done now. The efforts of Red

A Ministry of Food Recipe issued in October 1942 invited the nation to enjoy Mock Goose. The ingredients were 1½ lb. potatoes, 4 oz. grated cheese, two large cooking apples, ½ teaspoonful of dried sage, ¾ pint of vegetable stock, 1 tablespoonful flour, pepper and salt.

Dacre Street Church

Cross organisations can be supported and extended. It is surely a Christian duty to examine the facts without prejudice and try to discover the way God calls us to take.
Then we shall be troubled by the thought that to-day mankind is divided into two masses each striving to deprive the other of the very gifts God meant all to enjoy in common. Selfishness and recklessness in the use of God's gifts caused hunger before the war. And to-day we are far from heeding Jesus' injunction to be sons of Him who *is kind towards the unthankful and evil*, or from practising the spirit of Paul's words, *If thine enemy hunger, feed him.*
Thoughts like these may deprive us of the thoughtless happiness of those wrapped up in the satisfaction of their own wants. But we shall begin to discover the far deeper joys of co-operating with God's larger purposes for mankind. God will take the little offering of our love and imagination and sympathy and transform it by the mighty energies of His Spirit till it issues in a glorious harvest of peace and joy and love for the world.

A report of Alfred Bayly's sermon at Dacre Street Congregational Church, Morpeth, on Sunday, 11th October 1942, as printed the following Friday in the 'Morpeth Herald and Reporter'. 'The fruits of the earth, displayed with taste and skill in front of the pulpit, testified to a rich harvest'

The year ended with a party for Morpeth's Scouts an Guides at which Albert Bayly, in his capacity as Group Scou Master, gave a talk between items of fun. At a meeting of th Morpeth Ministers' Club, a look to the future beyond th dark days of world war led to a four-day World Churc Exhibition being staged in a local school in April 1943 It was open for just over seven hours daily and looked a the Church in action in the Pacific Islands, in war-torn Chin and village India, with a special place for Morpeth's ow missionary pioneer, Robert Morrison, and the world churc serving the local neighbourhoods on Tyneside. In openin the event, the Bishop of Newcastle, who had previousl served in Borneo, spoke of the restrictions and difficulties o staging such an exhibition in war-time. *If it had not been fo their secretary, the Rev. A. F. Bayly, he was sure they woul have bowed beneath those difficulties and given up Mr. Bayly kept them going. He had been a hard taskmaster rather like a very efficient headmaster.* The organisers me again in June and called on churches to study modern way of missionary service and to circulate books on the worl church.

The necessity for action to relieve Jewish persecution and t provide safety for refugees was discussed at length at Churc Meeting in July. The full extent of the holocaust was not ye ealised in England. In September, a meeting after Church ook the issue further under the title *The Jew our Brother.* n 1943, too, a Church Training Corps was begun and, hough numbers were small, some deep subjects were liscussed. At the first club session in September a RAF Pilot)fficer gave a talk on aircraft. They tackled the origins of the Quakers, how Methodism got its name, and the question *f Christ was alive to-day, would he join the Army?*

There was a Christian view of the war and however absorbing and pre-occupied, it was only an incident in that greater part, which was the great divine part of bringing the whole world to learn of the knowledge and love of God. They must refuse to be daunted by difficulties. After all, one of their leading characteristics was that the Christian view and the Christian way meant the overcoming of obstacles by the power of Christ. All concerned with that exhibition had been inspired by that supremely Christian principle and view of life.

From a report in the 'Morpeth Herald and Reporter' of the speech by the Bishop of Newcastle, Dr. Noel Hudson, at the World Church Exhibition

Sunday January 2nd 1944. The Session opened as usual with prayers from the Vigil. Mr. Bayly then read the log in the absence of Gordon Walker and Sidney Jefford, the Council members. As Alan Raitt was the only member present, the next Training Course was discussed, but not fixed. The collection for the L.M.S. was then taken. Alan read the Overseas Communique from China. Mr. Bayly then answered a question asked by Frank during the Training Session of Oct. 24. The question was "What is the attitude of Christianity to National Socialism, nationalism and Communism; and vice versa." First of all, it was explained that National Socialism held that man was made for the state, and that Adolf Hitler and the Aryans headed the state, according to themselves. Several passages in the Bible were looked up and the conclusion was drawn that the State's commands should be obeyed as far as conscience allowed. National Socialism held that all was for the state, and so clashed with Christ in that respect. Further passages were read by Alan from the Bible, and these showed that in the eyes of God all men were equal. Here again Christ clashes with National Socialism. In reply to the second part of the question, Mr. Bayly said that Hitler said little about Christianity in "Mein Kampf," but that Herr Rosenberg had quite a lot to say about it in his book, "The Myth of the 20th Century." Together with Dr. Ley and Goebbels, he held that God had sent Hitler as a second and greater Christ to lead Germany to its true place in the world. At first, Germany accepted National Socialism as a Christian organisation, but later the Barmen Conference spoke against it, and many priests were killed for their teachings. It was decided to hold no Club session until a good attendance at a Training Session was obtained. The session closed with prayers.

Albert F. Bayly 9.1.44.

*Wednesday January 12 ~~As usual, we opened with prayers from the~~ Mr. Herson and Alan played two games of chess, and Billy wrote a sketch for a concert which had been suggested.

Albert F. Bayly

Sunday January 9 As usual, we opened with prayers from the Vigil. The log was read, and the collection was taken. Alan then read the Overseas Communique. Mr. Bayly then answered the second half of Frank's question. Mr. Bayly explained that

Church Training Corps minutes - January 1944

:hat same year, his wife's interest in world needs was evident vhen she proposed the vote of thanks to a speaker on 'amine Relief at a well-attended Women's Conference in he Town Hall. At Harvest-tide, Albert Bayly again took a vorld view: *God wills that men should co-operate, and onsider one another, and that none should be deprived of he necessities of life.* His telling of a story about China called Christians to high sacrifice: in the Manse they were always eady to share their small rations with the needy, even if after erving a refugee or entertaining troops, the cupboard was are save for porridge oats. At the end of November, it was ppropriate that Albert Bayly should be the preacher at a Jnited Service for fellowship and prayer for the world-wide hurch at St. James'.

During a flood in China some Christians were able to save some food. They knew it might not carry them through till relief came. But other people had none. The Christians shared their stores with those in greater need though it meant probable starvation for themselves. Are we helping to create a world in which that will be a normal and natural thing to do?

From a report in the 'Morpeth Herald and Reporter' of Albert Bayly's Harvest Sermon, 1943

n 1945, the war ended with the capitulation of the Japanese ollowing the dropping of atomic bombs on Hiroshima and Nagasaki. On the 19th August, Albert Bayly preached at joint service of Presbyterians, Congregationalists and Methodists taking as his text Isaiah 25.7, concerned with an arlier deliverance of the tribes of Judah.

We pray for Christians everywhere, to-night. Let them look out upon a wider horizon, upon the Great Soul of the world.

From the 'Morpeth Herald and Reporter' quoting Albert Bayly's words at a United Service at the end of November 1943

They must give deep thanks to God for their deliverance. Before them lay new and brighter possibilities than known for a long time. At the moment they were chiefly hopes and possibilities. The veil was lifted, but not yet actually removed. They could see a light ahead, but not yet a light in which the whole world could freely rejoice.

They stood half-way through the Valley, with the darkest of war experiences behind. Treachery had failed, pride humbled and violence ceased. Just as the Prophet said in his own time great retribution had fallen on evil forces which plunged the world into misery. Physical power behind the Nazis and Imperialists of Japan had been broken, and their ambitions frustrated. They at least would not be able to trouble the peace of the world for years to come. All that brought a sense of relief from fear and anxiety. With it would go deeper and wider occasions for thanksgiving.

Modern war involved a race in destruction in which both sides were forced, even into more fearful and indiscriminate methods of slaughter. Within the last few weeks those had reached a climax of horror which had disturbed even the hardened war leaders. That slaughter had now stopped. That surely was a reason for thanksgiving which could only be fully expressed by dedicating themselves to put all heart and energy into making the peace which had now began real and firm.

They had spent £500,000,000 and untold energy and thought on perfecting a power which could unless they became a better people, easily destroy humanity. Were they prepared so to give themselves to God and let Him rule their life that He might fit them to use the new energies now released to enrich instead of destroying human life?

From Albert Bayly's sermon on 19th August 1945
as reported by the 'Morpeth Herald and Reporter'
His text was Isaiah 25, 7.

At Harvest, he compared the impossibility of feeding multitude with five loaves with the task of caring for th millions suffering the consequences of six years of world wa *Give these loaves to me, and God will do with them fo these people what you cannot do yourselves*, he paraphrase the words of Jesus in the Scriptures. In November, as th Chairman of the Personal Welfare Committee of Morpet Community Council of Social Service, Bayly was delighted a the opening of an *Over-60's* Club.

For the Triple Jubilee of the London Missionary Society, h wrote *Rejoice, O people, in the mounting years*. The thir verse was written at the request of Dr. F. A. Ironmonger f

Rejoice, O people, in the mounting years
Wherein God's mighty purposes unfold.
From age to age his righteous reign appears,
From land to land the love of Christ is told.
Rejoice, O people, in your glorious Lord,
Lift up your hearts in jubilant accord.

Rejoice, O people, in the years of old,
When prophets' glowing vision lit the way;
Till saint and martyr sped the venture bold,
And eager hearts awoke to greet the day.
Rejoice in God's glad messengers of peace,
Who bore the Saviour's gospel of release.

Rejoice, O people, in the deathless fame
Won by the saints whose labours blessed our land;
And those who wrought for love of Jesus' name
With art of builder's and of craftsman's hand.
Rejoice in him whose Spirit gave the skill
To work in loveliness his perfect will.

Rejoice, O people, in this living hour:
Low lies man's pride and human wisdom dies;
But on the Cross God's love reveals his power;
And from his waiting church new hopes arise.
Rejoice that, while the sin of man divides,
One Christian fellowship of love abides.

Rejoice, O people, in the days to be,
When o'er the strife of nations sounding clear,
Shall ring love's gracious song of victory,
To east and west his kingdom bringing near.
Rejoice, rejoice, his church on earth is one,
And binds the ransomed nations 'neath the sun.

Rejoice, O people, in that final day
When all the travail of creation ends;
Christ now attains his universal sway,
O'er heaven and earth his royal Word extends:
That Word proclaimed where saints and martyrs trod,
The glorious gospel of the blessed God.

Albert Bayly's final revision of the text

Talking in this way, we entered a sanctuary which immediately impressed me by its simple beauty. In shape it was two semi-circles, a larger and a smaller one, joined along a common diameter. The larger one was set with chairs, not pews, for the congregation. The smaller one formed a semi-circular apse, which contained the communion table and pulpit.
I studied first the frescoes on the sanctuary walls, and saw that they represented people of many nations bringing the glory and honour of the 21st century world into the city of God – a strikingly modern but beautiful city pictured on the walls of the apse.

From Albert Bayly's dream of the English Church forty years on from 1965

he Festival celebrating the 750th Anniversary of Lichfield ;athedral. It was noticed by Canon Cyril Taylor, who was ıstrumental in securing its inclusion in the Hymn-Book of ıe British Broadcasting Corporation to the tune

NORTHUMBRIA by Walter K. Stanton. It was perhap
Cyril Taylor's encouragement that set Albert Bayly firmly o
the road to a hymn-writing career. The hymn was sung to
tune by H. P. Chadwyck-Healey GAUDEAMUS IGITUR i
St. Paul's Cathedral, London on St. Cecilia's Day, 1950
Eric Thiman also set the text to the tune REJOICE, C
PEOPLE: and it has been sung to SONG I. This hym
provided the title for the first booklet of Bayly's hymns an
poems published in 1950.

Dacre Street Church and the former Presbyterian Church were sadly unable to agree a merger in 1973. The Dacre Street Church came to the end of its financial and human resources in 1977. The building now houses a Full Gospel Church in association with the Assemblies of God.

Morpeth said good-bye to the Baylys at Oliver's Café a
the end of July. It was recalled that Albert had written a
the pageants and scripts for young people's celebration
during his ministry there. The Assistant Church Secretar
emphasized how Albert's *worth, his pity, his catholicit*
of spirit, by precept and example, revealed him a grea
Christian. Other tributes described *his quiet, retiring natur*
as of the 'salt of the earth'. He never climbed into a positio
he could not fill; his integrity was one of his most out
standing qualities. Marjorie was acknowledged as one wh
had *taken her place in town life . . . most generous in he*
service to one and all. In his final year, Albert Bayly serve
as Chairman of the Durham and Northumberland Con
gregational Union.

During his years in the North-East, Albert Bayly foun
relaxation in the Cheviot Hills and his gift as a painter h
recaptured many of those scenes. He went back for holiday
again and again in later years.

Nonconformity came to Burnley in industrial Lancashire i
1850 with early meetings in a room above William Pate
shop. Here Mr. George Partington preached three tim
each Sunday and met the friends *for social prayer an*
religious conversation twice during the week. The work m
with *violent opposition from various and some unexpecte*
quarters and *through ignorance or malice, or both, th*
doctrines he preached were represented in a most horr
light. To make the mischief against him as effective
possible, pamphlets attacking Partington were distribute
widely. Eventually Bethesda Chapel was built in 1814, but i
1849 some members seceded to form Salem Chapel opene
in 1851. At that time in Burnley children aged eight were st
employed in the mills for six-and-a-half hours a day, with tw
hours' compulsory schooling added.

Grey are houses,
Dark are the mills,
Tall are the chimneys
By the great hills,
What is the pattern
Of life that you weave,
Christ and His people
In your Hollingreave?

Towards the end of the century, Salem founded a daught
church in Hollingreave Road and it was to this cause th
Albert Bayly was called in 1946. Like most towns

industrial Lancashire, Burnley, principally concerned with spinning and weaving, enjoyed mixed economic fortunes, but Bayly's sojourn there coincided with a period of rising prosperity. Bayly's Christmas greeting to his new church picks up the imagery of weaving.

With the help of his wife, Marjorie, a Young Wives' Fellowship was started for which he wrote a Hymn for Homemakers, *Lord of the home, Thine only Son,* sung to MELCOMBE: WARRINGTON, ILLSLEY (BISHOP), WESTFIELDS and VERMONT have also been used.

Lord of the home, Thine only Son
Received a mother's tender love;
And from an earthly father won
His vision of Thy home above.

Albert Bayly: The opening verse of 'A Hymn for Homemakers'

Burnley in the 1950s

If Christ were born in Burnley
This Christmas day,
This Christmas day;
I know not if the busy throng
Would bid Him stay,
Would bid Him stay.
But He might rest,
My heart's own guest,
Of praise and glory worthiest;
If Christ were born in Burnley.

Albert Bayly: The third verse of 'If Christ were born in Burnley'

Woven of courage,
Sacrifice, love,
Faith's shining vision
Sent from above;
Such is the pattern
Of life that we weave,
Christ and His servants
In our Hollingreave.

Hearts pure and humble,
Hands that are strong,
Deeds that are kindly,
Laughter and song;
Christ gave the pattern
Of life that we weave,
He and His comrades
In our Hollingreave.

Youth that is eager,
Age that is wise,
Manhood's full vigour,
Children's bright eyes;
These make a pattern
Of life that we weave,
Christ and his brethren
In our Hollingreave.

Home circles happy,
Work done with zest,
Play that is healthy,
Worship and rest;
Fair is the pattern
Of life that we weave,
Taught by the Master
In our Hollingreave.

Christ the great weaver
Stands by His loom,
Calls to endeavour
All in the room;
Glorious the pattern
Of life we shall weave,
Christ and His workers
In His Hollingreave.

'Hollingreave': Albert Bayly's Christmas Greeting in connection with Hollingreave Congregational Church

Cradled among the hills, Kendall Gale responded from childhood to their fascination and challenge. Pennine fells and moors were a fitting prepartion for a life spent among the still wilder hills of north Madagascar. . . . In 1928, Gale spent six months in the Marofotsy country, visited over three hundred and thirty villages and established forty-six new causes. In some villages, such as Andranomianatra, he visited every home. *I have since had a letter from the pastor telling me of the wonderful result of that visitation. This is the line of progress, if only I could multiply myself a hundredfold.*

From Albert Bayly's biography of 'Kendall Gale'

In 1947 Albert Bayly wrote for the London Missionary Society a commentary on the life of Kendall Gale entitled *A Whirlwind for Christ*, a forerunner of a full biography published some years later and, in doing so, perhaps was a little disappointed that his own two attempts to serve as a missionary came to nothing. On the first occasion that he offered he was turned down on medical grounds and on the second, having been selected for the job, it was eventually given to a local volunteer no doubt to save expense.

It was at Burnley that Bayly began hymnwriting in earnest. His *O Lord of every shining constellation* echoes his continuing interest in astronomy: *Lord, save Thy world; in bitter need* expresses concern at the misuse of science. *If Christ were born in Burnley* set to FAIRHOLM ROAD sucessfully achieves what many have tried, the marriage between the mystery of the Incarnation and the here-and-now of an industrial town. He began a series of hymns inspired by characters from the Old Testament which were to form a substantial section when his first collection was published in 1950.

Rulers of men, give ear!
Should you not justice know?
Will God your pleading hear,
While crime and cruelty grow?
Do justly,
Love mercy;
Walk humbly with your God.

Masters of wealth and trade;
All you for who men toil:
Think not to win God's aid,
If lies your commerce soil.
Do justly,
Love mercy;
Walk humbly with your God.

From 'Micah' (beginning 'What does the Lord require') written in January 1949 and sung to SHARPTHORNE and TYES CROSS

Must we, who over nature hold
Imperial sway,
See all its glories manifold
Soon pass away?
Must we, whose planes the skies ascend,
Whose cities steel and flame defend,
In universal carnage end
Our little day?

From 'Nahum' (beginning 'She was a city proudly strong') written in December 1949. Sung to GRAVETYE

Albert's years at Burnley were saddened by Marjorie's ill-health and death on 9th April 1948.

Bayly's interest in ecumenism was marked by a personal visit to Amsterdam for the great assembly of the World Council of Churches in 1948. This led to his writing, in advance of his going, *A Hymn for Amsterdam* beginning *With jubilant united strain* for which Eric Shave wrote the tune AMSTERDAM HYMN. The text was published in *The Christian World*. Albert also spent a week in Iona where he made a small contribution to the building of the Abbey by working on a beam for the refectory.

We greet the day which prophet eyes
Desired to see:
When round the world Thy sons arise,
Whom Christ made free.
From every nation, tongue and race,
We tell the story of Thy grace;
And lift our eyes to Jesus' face,
One family.

Verse 2 of 'A Hymn for Amsterdam' which begins 'With jubilant united strain'

An example of a hymn picking up modern imagery sung to REDHEAD No.46 (LAUS DEO) and STUTTGART (ST. ANTHOLIN) is *Lord thy kingdom bring triumphant.* He was to update it later in the light of one or two corrections of scientific fact, changes in technology and missionary strategy. Verse 6 for instance was omitted and verse 7 altered when the hymn was printed in 1982 as:

By the living voice of preacher,
By the skill of surgeon's hand
By the far borne broadcast tidings
Speaking peace from land to land.

Christ works through His Church now as in the first century. Not only when the Apostles healed a lame man, but in the modern missionary hospital. Not only when Stephen died, but in the witness of martyr churches in modern Europe. Not only when Peter accepted Gentiles into Christian fellowship, but when the Church to-day opposes anti-Semitism and the colour bar.

From 'What is Christ doing now?' an article by Albert Bayly in the 'Burnley Express and News' 27th March 1948

Lord, Thy Kingdom bring triumphant,
Give this world Thy liberty,
May Thy Spirit's strong compulsion
Rule our tides of energy:

Where the vessel cleaves the ocean,
Turns the wheel, and wings the plane;
Where the miner toils in darkness,
And the farmer sows the grain.

Consecrate Thy people's labour
At the airfield, mill and port;
With the gladness of Thy presence
Bless our homes and grace our sport.

Let Thy mercy and Thy wisdom
Rule our Courts and Parliament,
And to soldier, sage and scholar
May Thy light and truth be sent.

By the pioneer's endeavour,
By the word of printed page,
By the martyr's dying witness,
And Thy saints in every age:

When Thy Church in ancient wisdom
Keeps her sacramental rite;
Or on lonely mission station
Lifts the curtain of the night:

By the living voice of preacher,
By the skill of surgeon's hand,
By the ether borne evangel
Speaking peace from land to land:

Lord, Thy Kingdom bring triumphant,
Visit us this living hour;
Let Thy toiling, sinning children
See Thy Kingdom come in power.

The original version of 'Lord, Thy Kingdom bring triumphant'

Those who look below the surface find that Christianity still has its ancient power, and that the creative energy of God's Spirit is still at work in His Church. It has contributed the most healthy elements in the life of modern society, and it holds the seeds of new life for the world of to-morrow.
Christianity cannot identify itself with any existing social order, capitalist, Communist, or any other. It works as a creative, and sometimes a disruptive force within them all, for it represents an order and quality of life which do not belong only to this world, but derive from an eternal world beyond. To be a Christian is to hold the future in our hands in a sense which those whose hopes are only in this world can never do.

. . . . .

Christians, therefore, are not overwhelmed by the vast changes through which the world is passing, even by the rise and apparent triumph of evil forces. These have their day, but from the vantage ground of eternity, or even in the long perspective of time, their triumph is vain and empty. The quiet hidden forces of faith, goodness and love, planted in human hearts by the Divine Spirit, hold the promise of the future and the assurance of eternal victory.

Albert Bayly: From 'The Force of To-Morrow', an article published on 9th September 1950 in 'Burnley Express and News'

Other hymns were inspired by the Triple Jubilee of the Church Missionary Society; a visit to his old College Chapel, a trip to see the musical *Lilac Time* and the bi-centenary of the father of English hymn-writing Isaac Watts.

Working in Burnley on the other side of the town was a home missionary of the Congregational Church. The role played in that Church by home missionaries was originally to try and build up a new cause or down town church until it was sufficiently strong to be able to call a minister. This was the task of Grace Fountain labouring on the other side of Burnley to Albert. Grace had worked previously as a deaconess on the industrial estate near the Ford Motor complex at Dagenham in Essex. Unbeknown to him, she also attended the Amsterdam Conference and was equally interested in ecumenical affairs. Grace and Albert were thrown together from time to time as fellow-Congregationalists and in 1950 Albert was to take Grace as his second wife.

When in 1950, Bayly's first collection of hymns was privately published, a reviewer in the *Mansfield College Magazine*, J. F. Shepherd, saw the mantle of Edward Shillito (1872 - 1948), author of *Away with gloom, away with doubt*, to have descended on Bayly. He noticed that sometimes *Bayly appears to have been hampered by tradition, and thereby has lost something of force and originality*: but in general heralded Bayly's success in writing hymns for his own age. In his preface to the collection, Bayly acknowledged the help he had received from Bernard Manning in his venture of writing hymns.

Swanland Congregational Church, Humberside

When Albert's application for missionary service was turned down, he and Grace went to Swanland, a village not far from Hull on Humberside. It was a small country cause and most of the activities were organized either by Albert or Grace. While there he wrote the librettos for three cantatas: *The Divine Compassion; Look on the Fields* and *Song of Bethlehem* for the composer W. S. Lloyd-Webber, Principal of the London College of Music, organist of Westminster Methodist Central Hall and father of Andrew, whose musicals have become well-known across the world.

> God's House looks forth upon the quiet street,
> Across the pond where white swans proudly ride,
> A well-house, school and village sign, beside
> The pool, are clustered where the four roads meet.
> Amid such peaceful scenes once moved the feet
> Of One who, humbly born, was crucified;
> But rose that with His own He should abide;
> And all the powers of death and sin defeat.
>
> So long ago! So far away! But no;
> He walks to-day unseen our village ways:
> And when, at morn or eve, His people go
> To break the bread, or offer prayer and praise,
> Or hear the Christmas tidings told; then, lo!
> Their eyes are opened, and upon Him gaze.
>
> *Albert Bayly: Swanland*

Albert describes Swanland in a poem which was also set as an anthem by Lloyd-Webber under the title *He walks unseen.*

In 1956 the Baylys returned to Lancashire, at Eccleston, a fast-growing suburb of St. Helens. This town, famous for its glass making, enjoyed more prosperity than many Northern towns and in the 1950s and 1960s was trying actively to diversify its industrial base. The local paper for that period mentions salt-glazed stoneware, packaging materials, reinforced plastics, steel drums and chains, bricks and pre-cast concrete among the variety of industrial activities complementing glass-making.

Albert Bayly, this *quiet, modest person and a great visitor who walked tirelessly on his errands of mercy,* to quote a church member, was welcomed at an Induction Service on 13th November 1956: and at his first Communion Service, the Lord's Table was covered with a new set of table cloths made by a Mrs Moore, sure to add *beauty and dignity* to this sacred office.

During his time at Eccleston, the monthly minister's letter constantly challenged the congregation to look well beyond their own locality. He mentioned how much his wife and he had enjoyed over the years entertaining German students who came to Britain under a scheme sponsored by an organisation encouraging Anglo-German Educational Relations. *If you also can offer hospitality, write to . . .* He championed the work of the Commons and Footpath Preservation Society and the Council for the Preservation of Rural England. He sought aid for the victims of five cyclones which, in 1959, *caused immense damage and*

I have heard of churches like that closed tea-room, whose members would *rather resent any newcomer* wanting to join them! No doubt there was spiritual food for those inside, but the door was locked to any who would come in. What a travesty of a Church claiming to have a Gospel for mankind! – the Gospel of One who came to *seek and to save that which was lost.* True, it is equally fatal to have an open door and no Bread of Life within, the condition of a Church that offers nothing the world cannot give. An open door to a well furnished spiritual table is the condition of a healthy Church fulfilling its true mission. This depends very largely on the attitude and quality of life of all of us who are church members. The open door is the welcoming spirit ever eager to bring new friends into the fellowship, even going out of the way to seek them. We must have spiritual food in our Christian experience, ready to share with those who need what Christ has given us. Are we helping to provide the Church with this open door and well-furnished table?

From Albert Bayly's 'Church News-letter', September 1958

serious loss of life in Madagascar. He reminds readers, a road casualty figures peak that *the problem has mor aspects which most concern the Christian conscience an sense of responsibility as a citizen.* Whilst he highlight the peak accident times associated with driving after eve a little alcohol, he reminds drivers, cyclists and pedestrian alike that *we are all guilty at times of taking risks we hav no right to take, for they involve not only our own lif but the lives of others.* As a non-motorist he acknowledge that *reckless swerving and thoughtless crossing by pedestrian account for many accidents.* At election time, he stresse besides home issues, the need *for adequate steps to win bac African confidence in Nyasaland* (now Malawi) *and elsewher in Central Africa.* In 1960, he links the sins of *pride, hatre and selfishness which crucified God's love incarnate in Jesu* with those *producing in South Africa the bitter fruits o wrong racial relations and policy.* Writing of the Genev Disarmament Conference, he reminds yet again that *the sin which crucified Christ could bring a world disaster far vaste than the fall of Jerusalem.* In 1961, he set the Christma message and the cost of reconciling opposing forces *agains political tensions between East and West, between blac and white, walls in Berlin, strife and apartheid in Africa barriers to immigration in Britain.* But he reports positivel *on the entry of the Russian and other Orthodox Churche into the World Council of Churches,* and in 1962 quote extracts from the New Delhi Assembly of the Council A month later, he commends retiring collections for thos who had *suffered severely in the recent gales, Sheffield mos of all.*

Eccleston United Reformed Church, St. Helens

lbert Bayly always believed in Drama as a powerful ommunication tool and in April 1958 the Church Dramatic ociety presented *The Vigil* which looked at the questions aised for many minds by the Resurrection story. His ongregation included many intellectually gifted young eople with whom Albert established a special rapport with is willingness to think through the relationship between eligious and scientific knowledge. A newsletter in 1957 mphasizes that *Christians are needed who can give a reason or the faith that is in them even to a Communist and the ıan who thinks that science has discredited the Christian eligion.* He encouraged his church members to assist in isitation of the neighbourhood, and in the stewardship of alents. He helped them in their decision making to build sanctuary to supplement their all-purpose Hall and returned o St. Helens in May 1968 to cut the first turf.

t must have given Bayly quite a thrill to write in a newsetter which had mentioned several Third World countries hat a member of his Youth Fellowship had been led to nitiate a group to develop Christian Action in the town in uch matters as the race problems in South Africa. Bayly ook the initiative to encourage co-operation when the Reverend J. B. Woodbridge came as Vicar of Christ Church nd this began the ecumenical movement within Eccleston. ince then, Methodist and Roman Catholic churches have een established and there is an excellent record of cumenical partnership.

n the last years of his ministry at St. Helens, Albert Bayly vrote time and time again about the need for generous and ractical aid to assist refugees and the hungry and he eminded his congregation that there are times *when the Church must be active in protest against injustice and in fforts to end oppression.* As the *Food and Agricultural Organisation of the United Nations* challenged the nations o *free the world from hunger*, Bayly was encouraged by cumenical action through Christian Aid in these fields.

Vhilst at Eccleston, Bayly wrote many choruses for use t children's rallies organised by the London Missionary ociety. Hymns included one for Brownies at Leeds, urprisingly lacking city imagery, another for the twenty-first nniversary of the formation of Pilots nationally – there vas a group at the Eccleston Church – a text for a omprehensive school in Gateacre, Liverpool and a hymn for he Welsh celebration of Goodwill Day. The 350th Anniversary of the Authorised Version of the Bible prompted nother text.

At the end of June a world-wide effort was launched. sponsored by the United Nations and warmly supported by the Christian Church, to tackle with new vigour and resources the vast refugee problem. Welcoming the proclamation of this *World Refugee Year*, the Presidents of the World Council of Churches remind us that the plight of the homeless is the greatest human tragedy of our day. Christians can help to meet it through the Inter-Church Aid and Refugee service of the Churches, for which a worthy effort was made in St. Helens in May. We must press for adequate Government help not only for temporary relief, but to settle refugees permanently in new homes. Sick, infirm, handicapped and aged refugees may bring no economic benefit to the countries where they live, but our Christian Faith bids us bear one another's burdens. We cannot claim to be a truly Christian community unless we recognise this call and remember that it is not limited by national frontiers. Some of you have already shown a practical interest in the service of Agnew House London to elderly refugees. Let us take this and other work for refugees specially into our prayers, and do all we can to further the purpose of this World Refugee Year.

From Albert Bayly's 'Church News-letter', July 1959

Here is the place where God's hand shapes again
The living clay by cruel diseases spoiled;
And from the midst of injury and pain
Brings forth the comeliness that evil foiled.
Here heaven's wisdom uses human skill,
And Christ's compassion moves man's heart to care.
Here surgeon's knife and nurse's hands fulfil
God's plan the wasted tissues to repair.
Here death is challenged by resurgent life,
And love puts all despair and fear to rout:
As, disciplined and girded for their strife,
The conquerors of suffering go out.
This house of healing is a holy place,
Where dwells the Lord of life in kingly grace.

Albert Bayly: Burns and Plastic Unit,
Whiston Hospital, near St. Helens
September 1962

But lightly has time's restless hand been laid
Upon the gracious aspect of these walls,
Since that bright hour my memory recalls,
When first I saw their loveliness displayed
Upon a summer day. The beauty stayed;
But five and thirty years have filled these halls
With thronging younger life. Death's curtain falls
Between old friends who here their sojourn made.

Come! leave the past. Greet now a future fair.
Within these precincts soon there will arise
A college ampler than our fathers knew:
And we to whom they gave her torch to bear
Now trust with joy the flame that never dies
To younger hands, assured they will be true.

Albert Bayly: 'The Editor's Adieu'
written on giving up the editorship
of the 'Mansfield College Magazine'

Bayly was saddened that church mem
bers here as in his other churches spen
so much time in fund raising efforts. Th
real work of the Church could benefi
from direct sacrifical giving with th
time saved used in activity to sponso
personal spiritual growth and addition
Christian service. Nevertheless, the hym
written in 1962 for Eccleston Co
gregational Church Christmas Fair *Lor
of all good, our gifts we bring to The*
has proved a useful hymn sung at th
Offertory.

For several years Albert Bayly edite
the *Mansfield College Magazine*, whic
reviewed publications of ex-student
and generally provided news about them
In 1962, he was introduced to th
Queen Mother at the opening of the ne
residences at Mansfield College and
sonnet he had written for the occasio
was read.

After he preached at Thaxted Congr
gational Church on 12th August 196
Albert Bayly was invited unanimousl
by the Church Meeting to take up th
pastorate there. It was the first time i
his ministry that he had sought to b
placed in the south of the country bu
he wished to be near his own family an
that of his wife. The move was mad
against the trauma of the Cuban missil
crisis when the future of civilisation, a
we know it, hung in the balance.

He started work in Thaxted in December and by Februar
had written a hymn for the town to a tune BROOMFIEL
written by the Church's gifted organist Ruth Bennett an
harmonised by Joy Ashford of Saffron Walden Friend
School. It is printed as the preface to this chapter. Th
same month a Church Meeting accepted the suggestion o
starting a Church-to-Church correspondence with a Germa
congregation at Ratingen. Missionary endeavour remaine
at the forefront of Albert Bayly's concern. He reminded hi
new congregation constantly of their opportunities an
responsibilities to back the work and on 3rd April a ne
Pilots company was inaugurated.

The Baylys, both music lovers, found themselves living in a house occupied by the composer Gustav Holst from 1917 to 1925. On 5th July a plaque on the Manse wall was unveiled by Sir Adrian Boult in the presence of Holst's wife, his daughter Imogen, and Mrs. Ursula Vaughan Williams, widow of another great British composer. The Thaxted team provided a Morris dance and the Ancient Hymn translated from the Liturgy of St. James *From glory to glory advancing, we praise thee, O Lord* was sung to Holst's tune SHEEN. Sir Adrian read a sonnet especially written by Bayly for the occasion.

> Here was great music born: these walls have heard
> New strains spring living from creative thought.
> Here song and symphony took shape, and brought
> A strange new beauty to the hearts they stirred.
> The music-maker clothed the poet's word
> In robes of sound of finest texture wrought;
> Austerely strong, with mind's rich burden fraught;
> A mind no fear of human blame deterred.
>
> Such was the man we honour here; his name
> Adds glory to this town's rich heritage:
> And we, who now in Thaxted's story claim
> For Gustav Holst a fair, immortal page,
> Would learn from him, and all he gave, to aim
> With like integrity to serve our age.
>
> *Albert Bayly: 'The Steps, Thaxted'*

In the autumn of 1963, a series of occasional joint services with the Baptist Church began, with the object of releasing one or other of their ministers to lead worship in nearby village chapels and of creating a greater unity between the two churches. A Youth Group was launched for the Church's teenage young people. Bayly wrote a hymn for use at Baptism, *Father, Thy life-creating love* to be included in the 1964 publication of the *Rodborough Hymnbook* published to celebrate the 250th anniversary of the founding of Rodborough Tabernacle by George Whitefield. His *Praise and Thanks-giving* for the folk tune BUNESSAN was also first published in that book.

That Christmas, Albert Bayly was out carol singing in aid of Oxfam and this inspired Thaxted's own *Christmas Carol* written in January 1964. He went on to write texts for anthems, two of these, for Easter and Septuagesima, being set by W. S. Lloyd Webber. In August, the mountains which inspired the writer all his life, sparked off the text *Thy greatness is like mountains, Lord* to the Scottish Psalm tune INVERESK. It was first sung publicly by the Lima Choir of Glasgow at Stirling Festival Concert in May 1965. He contrasts the gentler countryside around Thaxted in a poem written in September 1964. That year he was caught up in the heady enthusiasm of Church union by 1980. The memorable Nottingham Conference stimulated Albert Bayly to look forty years ahead with a dream of the English Church at the beginning of the twenty-first century, which he presented at a meeting of the Congregational Churches in the St. Helens district.

> Thy greatness is like mountains, Lord;
> High as the peaks above:
> And like the waters deep and broad,
> The ocean of Thy love.
> As isles encompassed by the sea
> Enjoy a blest tranquillity,
> So, girdled by Thine arm,
> We dwell secure from harm.
>
> *The first verse of a hymn by Albert Bayly*

It is clear from the small attendance at Church on Sunday evenings that only a few people to-day value a second act of worship on similar lines to the morning service. We have to discover the most worth while way of using, as a Church, the Sunday evening period. We are making experiments this autumn by planning evenings with particular groups in view . . .

From Thaxted Congregational Church's Newsletter of September 1964

At the beginning of 1965, Bayly submitted two entries in the Free Church Choir Union competition for a hymn on *Worship* and *Most high and holy Lord*, with recommended tunes NORTH PETHERTON or DOWN AMPNEY won the Arthur Berridge award.

In March, Albert Bayly's letter to his Church was primarily concerned with Amnesty International of which he started a successful group in Thaxted. After listening to the London Missionary Society's sermon preached by the Rev. R. W. Hugh-Jones at the Congregational Union Assembly in Westminster Chapel in May 1965, Bayly was moved to write his hymn, entitled *Gethsemane, There is a garden whither Jesus came.*

The musical tradition of Thaxted was not simply related to its one-time resident Gustav Holst; its rich musical heritage was greatly enjoyed by Albert and Grace, and it was celebrated in a poem *Music in Thaxted.* That September, a further poem *Christ in Thaxted* found evidence of Albert's interest in field, factory and shop, as well as in music, the Word and in sacrament. He ends significantly:

> But when I meet in Thaxted
> A childlike loving heart, and see
> A humble, faithful life;
> I need not seek, for Christ seeks me,
> As once on man Divine compassion smiled,
> And sought us all incarnate in a Child.

Christians must lend a hand to every enterprise which can further the purpose of the Prince of Peace, and the aims set out in the U.N. Charter are aims we can fully endorse. *To save succeeding generations from the scourge of war . . . to reaffirm faith in fundamental human rights . . . to establish conditions under which justice and respect for the obligations arising from treaties . . . and international law can be maintained, and to promote social progress and better standards of life in larger freedom.*
We may differ on such methods as the use of force to achieve these ends, but over a wide field most Christians can find in the U.N. a powerful instrument to serve Christ's purpose to reconcile men and bring them into right relationship with one another. Let us use it to the full.

Albert Bayly in 'Thaxted Congregational Church Newsletter', November 1965

The same month, at the request of his sister, Dorothy, Albert wrote a hymn to celebrate the centenary of Dr. Barnardo's Homes *Father, whose great love enfolds.*

In October when the London Missionary Society became part of the Congregational Council for World Mission, and the concept of mission as a circle of giving and receiving was formalised, Albert wrote a hymn for the final Board Meeting of the old Society.

In his November newsletter to Church members, Albert Bayly recalled the League of Nations and commended its successor the United Nations. Albert Bayly owed much to books; he loved them and used them to the full. It was, therefore, unsurprising that he commended the importance of the *united effort of Christians to 'Feed the Minds' of millions with Christian literature.* He wrote no fewer than five hymns in the next few weeks. One was again on the theme of world-wide mission and another was intended for young parents. Three were commissioned, the first for the stone-laying ceremony in April at the new building of his sister's Baptist Church at Waterlooville, Hampshire; the second at the invitation of the President, Dr. Lee H. Bristol, Jr., and senior class of Westminster Choir College, Princeton, U.S.A. Later in life he was to be elected an Honorary Fellow of that College. The third commission came from the Saffron Walden and District Free Church Federal Council.

These parents need Thy wisdom's light,
Thy love within their heart,
Bless Thou their home, and for their task,
Thy Spirit's grace impart.

From 'Our Father, whose creative love'
ABBEY; BRADFIELD (ST. JOHN BAPTIST (Calkin)); SOUTHWELL (Irons); CONTEMPLATION (Ouseley); ST. HUGH (Hopkins)

Throughout his ministry it was Albert Bayly's practice to rise early and walk the streets and lanes of the town.

Morning in Thaxted brings a wealth of joy
As I greet the early workers . . .

To all early risers, then, Albert was no stranger. Perhaps he felt especially at ease at a time of the day when little is expected in small talk. Persistent and caring in his visitation, a few perhaps found his shyness and lack of light conversation a problem. Sometimes some of his country congregation found his erudite and carefully-prepared sermons a little beyond them. He was always trying to lead folk from where they were to a deeper understanding of the eternal mysteries and an awareness of dimensions of living beyond the bounds of their country town. The Sunday evening discussion services dealt with many aspects of Christian behaviour, including the Christian use of money. His newsletter reminded folk of *ill-health, ill-nourishment and ignorance in Zambia*, and shared the sorrow of the *tragedy at Aberfan* where a Welsh village lost a generation of children when a slag heap rolled down to overwhelm their school.

He urged in another minister's letter *We must tell men frankly that true 'progress' does not consist of such achievements as being able to travel faster and farther than before, or to kill more people more quickly, or even in having more of material aids to well-being, but in the growth in wisdom, charity, humility and faith* which make us truly children of God. He commended the work of the National Trust in a newsletter at holiday time.

The muse continued active through 1967, though pacifist Bayly was particularly saddened by the Six-Day War in the Middle East and was moved by the great achievement of

Among the serious issues which face us in the world to-day, the issue of human rights is not the least urgent. We look back with horror and shame on the days of religious persecution in our own land. But we may not realise that even now, in many parts of the world, great numbers of people suffer grave injustice and grievous penalties for expressing honestly held convictions on social and political as well as religious questions.
Some years ago nearly all the member states of the United Nations adopted the Universal Declaration of Human Rights which asserted the right of everyone *to freedom of thought, conscience and religion, to manifest his religion or belief in teaching, practice, worship and observance,* **with** *the right to freedom of opinion and expression.*
These rights are far from being observed in some parts of the world to-day. It is estimated that there are something like a million *prisoners of conscience* **in over forty countries now. A growing movement has arisen to mobilise public opinion in defence of the rights of such people. Amnesty International, founded only four years ago, has now some 400 groups in Britain and other lands.**

From 'The Minister's Letter' to Thaxted Congregational Church in March 1965

The National Trust in particular, needs support in trying to save such beauty from being lost for ever, such as the coast scenery which Enterprise Neptune is endeavouring to preserve.

Albert Bayly writing his Minister's letter for September 1966

Sir Francis Chichester in circling the globe single-handed in *Gypsy Moth IV*. The tune THAXTED inspired *O Lord of circling planets and all the stars in space*. There were hymns about faith firing human hearts, of the light given to humankind, along with one for marriage. He wrote an anthem text for the dedication of the Baptist Church at Waterlooville, now completed. The local surroundings inspired two sonnets and the release of a house martin back into the wild was compared to the wide realms to which Christians are called.

The days are past when precious time and energy were consumed in raising money for the Church. Money is given as readily and freely as it is needed for the work of the Church both local and universal.

From Albert Bayly's dream of the English Church in the 21st century. Written in 1965

O Lord of circling planets and all the stars in space,
Whose purpose through the ages our reason seeks to trace,
Thy greatness awes our spirit, Thy faithfulness our heart,
And far beyond conceiving, eternal Love Thou art.
No worship can be worthy, no words Thy praise complete;
Our life's devotion only can be an offering meet.

Then teach us in our living Thy will to know and do,
And keep us to Thy purpose both now and ever true:
By humbleness in service, endurance in our pain,
Help us for Thee to harvest life's wealth of precious grain.
So fit us by our labours, by suffering and love,
To be Thy servants always, here and in heaven above.

Before us shines the vision of all creation's goal,
When Christ will fully triumph and rule in love the whole;
When hatred, fear and falsehood and every deed of shame
Will yield to His dominion, and all men own His Name,
The Name of Him who, dying, endured the Cross and grave
To win us life eternal, and all creation save.

Albert Bayly

. . . Protest against the social wrongs has an honourable history as ancient as the Hebrew prophets – read Isaiah, Amos or Micah.
But protest alone is negative. If it springs from blind frustration, bitterness or self-interest it can damage instead of furthering a cause. Most of society's ills call for a more positive and constructive approach, inspired by moral conviction rather than resentment, with compassion as the deepest motive of all.

From Albert Bayly's News letter to Thaxted Congregational Church. June 1968

His 1968 writings included two carols and a Communion hymn *Creator and Provider of this our daily bread* written for an Anglican priest in charge of a church in Greater Manchester. Again he wrote with the tune THAXTED in mind. In September, his newsletter began with a reference to the Pope's Encyclical on birth control.

Thaxted, drawn by Albert Bayly

Peace, deep peace, the peace of God is here,
Peace of the hills that far as eye can see,
Like a still image of the ocean swell,
Rest in their ancient quietness.

Peace, God's peace, when spring awakes to life
Crocus and aconite and snowdrop pure,
Daphne and daffodil and violet,
First of earth's beauty blossoming.

Peace, deep peace, the song of moonlit brook,
Peace of night's quiet streets and lamplit homes,
Under the planets and far galaxies,
Glory of God's own majesty.

Peace, God's peace, where love indwells a home,
Making a sanctuary for tired souls;
Peace in communion of dear friends who share
Treasures of life in fellowship.

Peace, deep peace, within the House of God,
Voices and hearts in unison of praise,
While a skilled touch upon the organ keys
Summons all heaven's harmony.

Peace, deep peace, the peace of God is here,
As I have found this peace, so let me give
Peace to all troubled hearts, all strife-torn souls
Peace that is life's true blessedness.

Albert Bayly: 'Peace in Thaxted'

The Pope's Encyclical on birth-control has raised wider issues than those of its particular subject. For many Roman Catholics it has brought an acute conflict between the claims of Papal and Churchly authority on the one hand and of personal judgment and conscience on the other. We deeply sympathise with those who face these tensions, but they are bound to arise in a world which presents man with many new and challenging issues.

The controversy will not be without profit if it makes all Christians think more deeply about the nature of authority in religion. Some have found the ultimate authority in an infallible Pope, others in an infallible Book. For some the claims of tradition or the creeds have counted for most. Others have relied on the *Inner Light* and the judgment of the individual conscience.

But the only final authority is truth itself. A Christian is convinced that he has a vision of this in Jesus Christ. Other authorities can lead us toward the truth and help our mind and conscience to respond to it more fully. But as fallible human beings in a world in which new facts and possibilities are constantly coming to light we must always be ready to receive insight and revise our judgment.

From Albert Bayly's
Minister's letter
September 1968

In prisons dark and cold men lie,
Cut off from humankind;
While cruel torture violates
The flesh and sears the mind
Some toil long years in labour camps;
And others, exiled, mourn:
Like hunted beasts some men must hide,
From friends and kindred torn.

... Awake the conscience of mankind,
Let mercy temper might:
Release the prisoners of wrong
To liberty and light.

Lines from 'A Hymn for Prisoners of Conscience' beginning 'Our God, whose love in anger burns' sung to ST. MATTHEW

Growing unemployment, the hardship of rising prices, industrial tensions and break-downs, strife and suffering in Ireland, Pakistan and Vietnam . . . Christians alone cannot set right many of these things or end the suffering they involve. We have no special wisdom to decide the best steps to deal with most of them.
Nevertheless, we cannot with-draw from active concern about them, and leave them to *the secular world.* We must continually assert and help to keep before the community, the human values rooted in our Faith, lest they be forgotten in men's individual or sectional pursuit of material gain or selfish political aims.

From Albert Bayly's Church Newsletter July 1971

In 1969 he reported with joy that one delegate to the Uppsala gathering of the Fourth Assembly of the World Council of Churches suggested that the event marked *an end to old-style ecumenism, concerned mainly with the internal squabbles of the churches and with ecclesiastical engineering, and a beginning of the World Church primarily concerned with the needs of mankind outside the Church.* The Church Secretary's report makes it plain that Albert's ministry was to an ageing congregation with less and less able-bodied assistance for the work of the Church. A hymn for prisoners of conscience *O God whose love in anger burns* was inspired by the work of Amnesty International. During the Apollo II moon mission, his text *Great Lord of the Universe* included these lines:

But still beyond shine distant fires
From nebula and star . . .

. . .

With reverent minds to learn the laws
That order Nature's ways.

Psalm 8 and space exploration combined in a poem *Christmas in the space age.*

We look into the atom and find a world within
Of energies electric and particles that spin,
With power to build up all things, or destroy the universe;
And ours the choice of bringing a blessing or a curse.
We need, O God, the wisdom to use this power aright,
And make it serve your Kingdom, your realm of love and light.

Verse 2 of 'God's Universe' which begins 'We look into the atom and find a world within'

My sense of debt to our National Health Service has been deepened and my admiration for those who operated has grown. It has been very good to enjoy the kindly comradeship of fellow-patients and to see their cheerful courage.

From Albert Bayly's News letter to his congregation, June 1969

In 1969, Albert had the first of two eye operations, one of which was successful initially, the other less so. The 350th anniversary of the sailing of the Pilgrim Fathers provided a text calling us to adventure. *From age to age God summons men* begins with the call of Abraham. In July 1970, he wrote the text *Joy wings to God our song.* Just before Bayly's retirement at the end of 1972, he was overjoyed that within the time of his active ministry at least the union of his own church with the Presbyterians had come about and it was his privilege to conduct as his last service the dedication of Elders.

The name *Reformed* . . . speaks of our common heritage in the Reformed tradition which derived from John Calvin rather than from Martin Luther, . . . it was originally, and is still legally *Thaxted Old Independent Meeting.*

Albert Bayly on Thaxted Congregational Church becoming Thaxted United Reformed Church

Grace and Albert made their retirement home in the Springfield district of Chelmsford and shared in the life of Christ Church United Reformed Church, where Albert was appointed an honorary Associate Minister. Through retirement years, the Baylys were grateful to find that Chelmsford too had a rich musical life. Bayly continued to write hymns. The Centenary of the Leprosy Mission, Dr. Coggan's sermon at his enthronement as Archbishop of Canterbury and Nigel Calder's broadcast on his book *The Key to the Universe* sparked off texts, the last *With reverence and wonder we view your work, O God* once more with THAXTED in mind. His hymn written in June 1975 *In the power of God's own spirit* was sung to Cyril Taylor's THE SPAIN at the 1977 Hymn Society Conference Act of Praise in Salisbury Cathedral. Albert Bayly was honoured to become an Honorary Vice-President of the Hymn Society of Great Britain and Ireland.

Christianity . . . has a credible message to give in an age of immense growth in scientific knowledge, and it has much of importance to say to a world transformed by technical achievements and divided by extremes of wealth and poverty.

Albert Bayly, writing in his Church Newsletter of November 1972

Throughout his life Bayly had never owned a car and carried out the whole of his ministry on foot or by public transport. He was always a keen observer of the natural scene with a love of flowers, trees and the vistas of the countryside.

When Albert was in his mid-seventies he suffered a haemorrhage in his good eye. This led to near blindness and he gave up his preaching. This loss of sight was a great blow to one who loved books and the countryside so passionately. He was sustained by Grace who read to him extensively.

We ought, surely, to be more conscious now of our interdependence as members of society which, like the Church, is a body in which, as Paul wrote, 'The eye cannot say to the hand, *I do not need you.*' We can too easily take for granted the work of sections of the community with which we do not come into daily contact.

From Albert Bayly's Church Newsletter of March 1972 discussing the current industrial crisis.

'The Churches', added my guide, 'really began to be space-conscious about 1975. As man has explored more of the universe, many Christians have become more fully aware of the majesty and vastness of God's handiwork. Many Church Youth Groups now have a simple observatory, through which young people are introduced to a first-hand knowledge of the cosmos, and learn to discuss the issues which this raises for Christian thought.'

From Albert Bayly's dream of the English Church in the 21st century. The text was written in 1965

The hours are long, each day seems like a year;
we grope our way through corridor and ward.
Nurses and surgeon through a mist appear,
holding the promise of clear sight restored.

The sun in shining on the cricket ground
where Essex and Glamorgan fight it out,
and through an open window floats the sound
of batsman's stroke and keen spectator's shout.

The tide of active life around us flows,
but we are like an island off the shore:
and only one who waits here fully knows
the joy when loved ones enter at the door.

So pass the days until God's healing power
through human skill and care brings freedom's hour.

Albert Bayly: 'Eye Ward, Chelmsford and Essex Hospital'

Teach me, O Father, faith that leaps
To venture boldly at your call;
A faith that firmly stands the shock
When life's disasters fall;
A faith that trusts your love and power
To meet life's challenge every hour.

Albert Bayly: The opening verse of 'Double Trinity'

He continued to tend both the garden and an allotment and to busy himself in the training of lay preachers. A hymn text entitled *Double Trinity* beginning *Teach me, O Father, faith that leaps* and a poem *The Inner Vision* date from these difficult days. He tells the story of some restoration of sight in the poem *Eye Ward, Chelmsford and Essex Hospital.* With this recovery came the last burst of hymns and poems, including *Testimony*, a hymn beginning *Jesus Christ, I love and serve* and the poem *Oberammergau 1980.* He submitted an entry for Anglia Television's Easter Hymn Competition *Life is born of death to-day* to Michael Dawney's tune MARNHULL, and this was transmitted as a runner-up.

Over the years, Albert and Grace were regular attenders of the Conferences of The Hymn Society of Great Britain and Ireland. They were present at Chichester in July 1984. The day following its conclusion Albert suffered a heart attack and died a few hours later. In the words of his poem *Birth from Birth* he had lived

In tireless service to your people's needs,
Until my tasks for you, my Lord, are done.

Now he is discovering the new tasks of the life beyond mortality.

We count you now among the pioneers;
For you, dear friend, had reached this land before us,
Before the dawn of our explosive years.

You staked no claim, issued no manifesto;
And yet the songs you sang with quiet voice
Pointed the way we knew we had to go.

Perhaps you learned from us, as we from you;
Your talent flowered and fruited in old age,
In sharpened language and in rhythms new.

Therefore we come to lay our wreath of words
Not on your grave but where a rose is planted,
Saying: *Rejoice – the glory is the Lord's!*

Fred Pratt Green: 'In memory of Albert Bayly'

'And now', he said, 'look there'. He pressed a button, and a screen swung out on the opposite side of the apse. 'When we come here to worship' he continued, 'we are able, thanks to science, to see, projected on that screen, the world into which our Lord calls us to go, the hungry crowds waiting to be fed with material and spiritual bread. We see the Church at work and worship in many lands, seeking to meet the world's need. We see the men and women for whom Christ died to whom He is waiting to speak, the sufferers He is waiting to heal, the many sided life of men He strives to enter and rule.'

From Albert Bayly's dream of the English Church in the 21st century. The text was written in 1965

Tune 'Thaxted' by Gustav Holst

INDEX OF PERSONS

An * against a name indicates an entry in the corresponding index in *Hymnwriters* I

Aaron 26
Abbott, Thomas Easthoe 92
Ackermann, Rudolph 91
Adam 20
Adam of St. Victor 122
Adams, Mr. 87
Addison, Joseph 119, *
Aesop 11, 12
Aikin, John 90
Akenside, Mark 119
Alcaeus 73
Alford, Henry 121
Ambrose, Saint 121, 126, *
Amos 176
Anne, Queen 39
Ariosto 112
Arthur, Robert 12
Ashford, Joy 146, 172
Aston, Joseph 73, 89
Atkin, Mr. 87
Atkins, C Leslie 153
Bach, Johannes Sebastian 153
Bacon, Roger 17, 18
Bailey, John 77
Baird, Captain 11
Baker, Henry Loraine 122
Baker, Henry Williams 120-145,*
Baker, Jessy 133, 138
Baker, Louisa Anne, (née Williams) 122
Ball, Hannah 104
Ballantyne, James 90
Barbauld, Anna 97
Barber, J 74
Barber, Thomas 96
Barham, Thomas Foster 27, 90
Barnardo, Thomas John 174
Barrie, James 152
Barton, Bernard 90
Bayly, Albert Frederick 146-181
Bayly, Dorothy 174
Bayly, Grace (née Fountain) 168, 169, 179, 180
Bayly, Marjorie (née Shilston) 154, 155, 164-166
Bayly, Mr. (Albert's father) 147
Bayly, Mrs. (Albert's mother) 147
Beddoes, Thomas 28
Beddome, Benjamin 97
Bede, The Venerable 122
Beecham, John 112
Beethoven, Ludvig van 76
Behmen, Jacob (*See* Boeheme)
Bennet, Benjamin 2
Bennett, George 81, 88, 89, 104, 105, 115
Bennett, J C 149 150
Bennett, Ruth 146, 172
Bernard of Clai(r)vaux, Saint 121
Bernard of Cluny 121, 122
Berridge, Arthur 174
Bickersteth, Edward 83, 93, 107
Bickersteth, Edward Henry 142, *
Birchall, J 107
Bird, Charles Smith 91
Blackley, Mary 58, 59, 61, 62, 76, 105
Blackmore, Richard 59, 60
Blackwell, John 99
Blair, Robert 59, 60, 73
Blew, W J 127
Bloxam, J R 122
Boeheme, Jacob 32
Boleslas 86, 87
Bogaris, King Michael 85
Bonaparte, Napoleon 73, 79, *
Bonnor, Letitia 122
Borziwog, Duke 86
Boult, Adrian 173
Bourne, George 72
Bowman, Thomas 22
Bowyer, W 76
Brady, Nicholas 125
Brameld, Mr. 62
Brewer, Samuel 16
Bridson, John 11
Bristol, Lee H Junior 175
Brookes, Captain 92
Brooks, N C 112
Brougham, Henry 107
Brown, James Baldwin 92
Browne, Moses 20
Browne, Simon 97
Buchanan, Claudius 50
Bull, William, 27, 32, 35, 39, 40, 42, 43, 46, 53-55, *
Bullock, William 129
Bunbury, H 93
Bunyan, John 24, 54, 115
Burke, Edmund 64
Burns, Robert 69, 92, 119, *
Butler, Samuel 119
Byron, George Gordon 119, *
Cagogan, W B 27
Cain 26, 109
Calder, Nigel 179
Calvin, John 42, 51, 179
Campbell, Thomas 119
Candasa of Mosambique 94
Candasa (slave daughter of Candasa) 94
Carey, William 50, 88, 89, *
Carey, William Paulet 74, 93
Carolina (slave daughter of Candasa) 94

Caroline of Brunswick, Queen 86
Catlett, Betsy 27, 42, 44, 54, 56
Catlett, George 27
Catlett, John 15
Catlett, Mary 3, 4, 7-9, 11-18, 22, 24, 27, 40, 52, 53, 56
Catlett, Mr. (Mary's father) 3, 8, 27
Catlett, Mrs. (Mary's mother) 3, 8
Cato, Marcius Porcius 11
Cavanagh, Owen 11
Cawthorne, Mary 24, 28, 44, 54, *
Cecil, Richard 43
Cennick, John 59
Chadwyck-Healey, H P 164
Chaloner, Robert 99
Chamberlain, Neville 156
Chandler, John 127
Chantrey, Francis Leggatt 74
Charles II, King 108, *
Chatterton, Thomas 91
Chaucer, Geoffrey 119, *
Chichester, Francis, 176
Chow, Mr. 4, 5
Chudleigh, F W 157
Churchill, Charles 119, *
Clarendon, Edward Hyde 108, 109
Clarke, Adam 88
Clarke, Charles Cowden 93
Clunie, Alexander 16
Cobbett, William 76
Coffin, James 54
Coffin, Mrs. 54
Coggan, Donald 179
Colebrook, Lt. Col. 50
Coleridge, Samuel Taylor 103, 119
Collins, William 119
Collyer, William Bengo 78
Conder, Josiah 93
Cooke, Mr. 87
Cooper, Anthony Ashley 2, 3
Corrigal, Andrew 11
Cotterill, Thomas 83-85, 88, 126
Cowley, Abraham 119
Cowper, John 56, *
Cowper, William 20, 21, 24, 27, 32, 40, 43, 44,48, 53, 54, 56, 57, 94, 96, 97, 104, 119, 132, *
Crabbe, George 119
Cranmer, Thomas 127
Crashaw, Richard 119
Creed, Thomas 11
Cresson, Elliott 94
Crocker, Charles 93
Croly, George 94
Cromek, Robert Hartley 92
Crotch, William 127, *
Cumberland, Mr. 70
Cunningham, Eliza 50
Cunningham, John William 94
Cyrillus 85
Daniel 26, 29, *
Daniel, George 93
Daniels, Bebe 152
Dante Alighieri 108, 112
Darling, R S 151
Dartmouth, Earl of (*See* Legge)
Davenport, Richard Alfred 91, 92
David, King 121
Davies, David 78
Davies, E O 149
Dawney, Michael 180
Dawson, Jane (née Flower) 44
Denham, John 119
Denton, W 127, 128
Deutero-Isaiah 29, 176
Dix, John 91
Doddridge, Philip 97, *
Doloribus, Josephus 50
Donne, John 119, *
Drahomira 86
Drake, Mr. 87
Drummond, Robert Hay 19
Dryden, John 119
Duncan, Captain 9
Dykes, John Bacchus 136-137, 144, *
Edgar, King 123
Edwards, Letitia (née Bonnor) 122
Edwards,W 122
Elijah 26
Elisha 26
Ellerton, John 107, 108, 140, *
Ellis, William 91
Ephrem (Ephraim) Syrus 121
Esau 26
Estlin, John Bishop 91
Eva (slave daughter of Candasa) 94
Evans S A 153
Everett, James 87, 93
Ezekiel 29
Ezra 95
Falkland, Lord 108, 109
Fitzwilliam, Earl (*See* Wentworth-Fitzwilliam)
Flinders, Captain 102
Flower, Jane 44
Ford, Dr. 50
Forster, John 91
Fortunatus, Venantius 121, *
Foster, Elizabeth (née Montgomery) 86, 101
Foster, Henry 43
Foster, John 27
Fountain, Grace 168, 169, 179, 180
Francis Xavier, Saint 122
Gale, Kendall 166
Gales, Anne 66, 102, 112, 113, 118
Gales, Joseph 64, 66
Gales, Sarah 90, 102, 113, 115, 118
Galvani, Luigi 74
Gambado, Geoffrey 93
Gandhi, Mohandas 153

Gardiner, William 76, 91
Gariboldi, Guiseppe 131
Garrick, David 50
Garrick, Eva Maria 50
Gauntlett, Henry J 127, 132
George III, King 52, 58, 64, 76
Gerhardt, Paul 129
Gibbons, Orlando 129
Gideon 26
Goldsmith, Oliver 119
Gordons of Haddo 124
Goss, John 127
Gower, John 119
Gray, A Herbert 154
Gray, Thomas 44, 119
Green, F Pratt 181
Gregory the Great, Saint 121
Grimshaw, William 54
Grinfield, Thomas 91
Habakkuk 53, 121, *
Hadfield, George 91, 94
Hall, Samuel Carter 92
Hamilton, Mr. 61, 62
Handel, George Frederick 47-49
Hannah 26, 121, *
Hannibal 17
Harrison, Christopher Robert 128, 143
Harrison, W 62
Harrison, Reverend (brother of Christopher) 121
Harrison, Mr. (son of Christopher) 121
Harvey, Lilian 152
Havergal, W H 127
Haweis, T 1, 20
Hawkes, Thomas (Carpenter) 16
Hawksmoor, Nicholas 39
Haydn, F Joseph 76
Hayward, Mr. 16
Heber, Reginald 29, 97, 122, 126,*
Hemans, Felicia Dorothea 104, *
Herbert, George 119, 122, 136, *
Hetherington, William Maxwell 93
Hodgson, Francis 91
Hodgson, Rowland 115
Hofland, Barbara (née Hoole) 74, 75, 112
Hofland, Thomas Christopher 74
Holland, John 99, 118
Holst, Gustav 173, 181
Holst, Mrs. (wife of Gustav) 173
Hoole, Barbara 74, 75, 112
Horne, George 112
Hopkins, John 125
How, William Walsham 128, 140, 141, *
Howard, John 92
Hughes, E R 150
Hugh-Jones, R W 174
Huie, James 113, 115
Hunt, Joshua 61, 63
Hunt, Mrs. (wife of Joshua) 63
Hunter, Joseph 92
Hunter, William 50
Huntingdon, Susan 104, 105
Huntingford, George William 122, 128, 131
Illidge, Thomas Henry 92, 93
Illingworth, William 115
Ingham, B 19
Ironmonger, F A 162
Isaacs, J 147, 162
Isaiah 29, 176
Jackson, John 102
Jackson, Thomas 92
Jacob 26
Jarrett, Thomas 50
Jenner, Henry 130
Jennings, David 15
Jeremiah 29
Jinnotjee (slave daughter to Candasa) 94
Job 29
Jones, Thomas 25
Jonah 29
John 32, 121, *
Jonson, Ben 119
Joseph 5
Joshua 29
Keble, John 122, 129, 135, *
Keene, Mr. 23
Ken, Thomas 92, 122, *
Kennedy, Rann 92
Kennicott, Benjamin 42, 43
King, J W 65, 73
Kitchener, Horatio Herbert 147
Langland, William 119
Langton, S 99
Langton, Mr. 10
Lardner, Dionysius 112
La Trobe, John Antes 90
La Trobe, Peter 113
Lazarus 55, 113
Lees, William 10
Legge, William 19-22, 24, 27, 32, *
Le Noir, Elizabeth Anne 90
Lewis, Job 9, 16
Lewis, Nicholas 59
Lindsay, Coutts 142
Lloyd Webber, Andrew 169
Lloyd Webber, W S 169
Lockwood, Mr. 61
Lot 26
Lot's wife 26
Lowde, F C F 121
Lucellus 46
Ludomilla 86
Luke 32, 33, *
Luther, Martin 179, *
Lyall, William Herle 131, 143
Lydgate, John 119
Lyttleton, William Henry 93
Maberley, Thomas Astley 127, 131, 143
Mack, General 73

Madan, Martin 43
Manesty, Joseph 3, 7, 16
Manning, Bernard 168
Marconi, Guglielmo 147
Mark, Saint 32
Martha 33
Martin, E Neville 152
Marvell, Andrew 119
Mary, Saint 121
Mary (sister of Martha) 33
Matthew, Saint 32
Medhurst, Miss 17-19, 23, 24
Methodius 85
Micah 166, 176
Michael III, Emperor 85
Midian 116
Middleton, Charles 53
Middleton, Lady 53
Milman, Henry 136
Milner, Reverend Dr. 99
Milton, John 119
Milton, Lord 99
Miriam 121
Monk, William Henry 136, 137, 139,144
Montgomery, Elizabeth 86, 101
Montgomery, Ignatius 59, 113
Montgomery, James 58-119
Montgomery, John 59, 61, 62, 76, 105
Montgomery, Mary (née Blackley) 59, 61, 62,76, 105
Montgomery, Mr. (James's great grandfather) 59
Montgomery, Robert 59, 113, 118
Moore, Thomas 119
Moore, Mr. 113
Moore, Mrs 169
More, Hannah 50, 51, 53
Morrison, Robert 160
Moses 32, 121
Mozart, Wolfgang A 76
Murray, Francis Henry 123, 126-128, 139
Nahum 166
Nanson, Edward 76
Navarro, Ramon 152
Naylor, Benjamin 67
Neale, J M 127, 128
Nehemiah 29
Neele, Henry 92
Newman, J H 122
Newton, Captain (John's father) 1-5, 7
Newton, John xii, 1-57, 97, 104, 132
Newton, Mary (née Catlett) 15, 29
Newton, Mrs (John's mother) 2
Newton, Mrs (John's stepmother) 2
Noah 87
Noel, Caroline 139-141
Noel, Gerard T 141
Novello, Alfred 93
Oakeley, Frederick 132
Occum, Samion 24
Okeley, Francis 19
Otto I 87
Ouseley, Frederick A Gore 136
Padfield, John 133
Paine, Thomas 64
Parker, Hugh 99
Parnaby, Henry 149
Parnell, Thomas 119
Partington, George 164
Pates, William 164
Paul 32, 42, *
Pearson, George 64, 65
Peel, Robert 112
Peill, Arthur 153
Peter the Apostle 28, 167, *
Pickering, William 93
Pitman, Isaac 115
Pitt, William 43
Pollock, John 43
Pooroosh, Ram 50
Pope, Alexander 74, 119, *
Positive, Paul 68
Pott, Francis 129
Powley, Matthew 28
Powley, Susanna (née Unwin) 24, 28, *
Pratt, Josiah 90
Pringle, Thomas 92, 104, 105
Prior, Matthew 119
Pulling, William 128, 131
Quarles, Francis 119
Raffles, Thomas 93, 107
Raikes, Robert 104
Rameses II, Pharaoh 121
Reynolds, Richard 80, 81
Rhodes, Ebenezer 64, 65, 76
Richards, William Upton 128, 131-133, 143
Ridgeley, Mr. 42
Ring, Thomas 54
Ring, Mrs 54
Robbins, George Henry 90
Roberts, Samuel 81,112, 115
Rogers, Jacob 19
Romaine, Reverend 41
Rose, William 28
Ryan, Richard 90
Ryland, John 22
Sackville, Thomas 119
Saint, H B 151
Sallust, Gaius Sallustius Crispus 46
Samson 26
Samuel 29, 121
Santeuil, Jean Baptiste de 129
Saphira (slave daughter of Candesa) 94
Schweitzer, Albert 153
Scott, Captain 24, 25
Scott, Giles Gilbert 122, *
Scott, Thomas 29, 38, 39
Scott, Walter 108, 119, *
Sedulius, Caelius 122, *

Selbie, W B 151, 152
Shaftesbury, Earl of (*See* Cooper)
Shakespeare, William 119
Shearer, Norma 152
Sheppard, John 90
Shepperd, J F 168
Shillito, Edward 168
Shilston, Marjorie 154, 155, 165-166
Shilston, R D 154
Shilston, Mrs. (Marjorie's mother) 154
Shoberl, Frederic 90
Shore, Miss 78
Shore, Mr. 87
Silvertongue, Gabriel 64, 68
Simeon 121
Singha, Shoran S 154
Skelton, John 119
Smedley, Edward 94
Smith, Betsy (née Catlett) 27, 42, 44, 54, 56
Smith, John Pye 67, 68
Smith, Richard 118
Smith, Mr. (Betsy's husband) 56
Söderblom, Dr. 149
Solomon 26, 29, 31
Sophia, Saint 32
Southey, Robert 94, 119
Spangenberg, Augustus Gottlieb 59
Spencer, Thomas 76, 77
Spenser, Edmund 119, *
Stainer, John 134, 136-138
Stancliffe, Dr. 74
Stanton, Walter K 164
Starlight, Jonathan 64, 65
Stephen 167
Sternhold, T 125
Stoymirus 86
Street, George Edmund 124, 135
Sullivan, Arthur 140-142, *
Surrey, Howard Henry 119
Sylvester, Charles 64, 74
Symonds, Joshua 24
Talbot, W 27
Talbot, Mrs. 27
Tasso, Torquato 112
Tate, Nahum 125
Taylor, Cyril 163, 164, 179
Taylor, Jane 97
Tell, William 72
Thomas de Celano 122, *
Thompson, Gordon 102
Thomson, James 119
Thornton, John 21, 22
Timothy 42
Toplady, Augustus 35, 97, 122
Trench, Richard Cheneuix 134
True, Thomas 11
Turner, Hannah 63
Turretin, Mr. 42
Tyerman, Daniel 104, 105
Tymby, H B 99
Unwin, Mary (née Cawthorne) 24, 28, 44, 54, *
Unwin, Susanna 24, 28, *
Unwin, William 44
Varah, Chad 94
Vaughan, Henry 122
Vaughan Williams, Ursula 173
Veigel, Eva Maria 50
Victor Emmanuel, King 131
Victoria, Queen 136
Vydyunath 50
Waller, Edmund 119
Ward, P 128, 131, 143
Ward, Thomas Asline 74-76, 79
Warnech, H L 94
Watts, Isaac 53, 78, 98, 168, *
Webb, Reverend 134
Webber, Andrew Lloyd 169
Webber, W S Lloyd 169
Wenceslaus, King 86
Wentworth-Fitzwilliam, William 62
Wesley, Charles 22, 50, 92, 97, 122, *
Wesley, John 19, 22, 41, 92, *
West, Daniel 22, 23, 27, 28
West, Mrs. 23
White, G Cosby 127, 128, 139
Whitefield, George 16, 22, 23, 41
Whitford, Mr. 17, 28
Wilberforce, William 43, 44, 53, *
Wilferforce, William (uncle to the above) 22
Wilberforce, Mrs. (née Thornton) 22
Wilkins, John Murray 123, 131, 143
Wilkinson, James 69
William the Conqueror, King 39
Williams, Derek 147
Williams, Louisa Anne 122
Williams, William 122
Williamson, James 97
Wills, Iona 152
Wilson, Dorothy 150
Winkelried, Arnold de 108
Wisner, Benjamin 104, 105
Wither(s), George 119, 129
Witsius 42
Woodbridge, J B 171
Woodford, J R 128, 129
Wordsworth, Christopher 134, *
Wordsworth, William 119, *
Wormwell, Henry 68
Wratislaus 86
Wratislaus 86
Wren, Brian 148
Wren, Christopher 39, *
Xavier, St. Francis 122
Young, Edward 119
Zaccheus 33
Zacharias 121
Zechariah 29

INDEX OF PLACE NAMES

An * against a name indicates an entry in the corresponding index in *Hymnwriters* I

Aberdeen 113
Aberfan 175
Aberystwyth 104
Africa 5, 7-13, 45, 47, 48, 78, 92, 94, 95, 98, 104, 105, 108, 153, 157, 170
Alexandria 70
Alicante 2
Alloway 69
Alps 17, *
America 72, 112
Amsterdam 166-168
Angola 45
Antartica 155
Antigua 8, 44, 78
Antrim, County 59
Argyllshire 139
Arrow, River 123
Ashington 155
Asia 46, 50
Atlantic Ocean 46, 58, 112
Australia 92, 104
Austria 70, 73, 110
Ayr 113
Ballykenny 59
Baltimore 112
Bannockburn 69
Barbadoes 59, 113
Barnwell 122
Barrackpore 50, *
Bath 53, 113, 119, *
Beckenham 28
Bedford 24, 53, *
Bedfordshire 19, 24, 53
Bedlington 155
Belfast 66
Belgium 159
Bengal 46, 50, *
Berkeley 91
Berlin 170
Berwick 24
Bethelsdorf 102
Bethany 117
Bethlehem 169
Bexhill-on-Sea 147
Birmingham 92, 126
Blackburn 93
Blackfriars 41
Blackheath 20, 101
Black Rock (Liverpool) 9
Blenheim 94
Bohemia 86
Bonanoes 9, 11
Borneo 160
Boston (Mass.) 104, 105
Bow 154
Bradford 113
Brasil 45
Brecon Beacons 104
Brighton 90, 115
Bristol 53, 79, 91, *
Builth Wells 121
Bulgaria 85
Burnley 164-168
Calcutta 126
Cambridge 123, 130, 139
Cana 132
Canada 129, 147, 149
Cancer, Tropic of 13
Canterbury, 126, 135
Cape Lopez 5
Cape Mount 47
Cape of Good Hope 78, 94
Cape Saint Vincent 9
Cape Town 108
Carlisle 124
Cawston 22
Charlestown 8
Chatham 3, 4, 7, 13, 16
Chelmsford 179
Chelsea 127
Chepstow 104
Cheshire 79
Cheviot Hills 164
Chichester 93, 180
China 104, 115, 116, 150, 157, 160, 161
Chislehurst 123, 127, 139
Clairvaux 121
Cleveland 139
Clifton 25
Clyde, Firth of 59
Codrington 28
Colchester 122
Cologne 129
Columbia 131
Constance 149
Constantinople 85
Copenhagen 70
Cornwall 54, 130, 147, *
Cottenham 24
Cowslip Green 51, 53
Cuba 172
Cuckfield 127, 131
Cullercoats 151, 153
Cumbrae 127, 139
Czechoslovakia 157
Dagenham 168

Danish West India Islands 78
Darford 63
Darlington 92
Deptford 1
Derbyshire 65, 101
Doncaster 66
Doon, Loch 69
Dover 53
Dundee 113
Dunedin 130
Durham 136
Durham, County 164
East Indies 4
Eastnor 128
Eccleston 169-171
Edinburgh 76, 92, 113, 115
Egypt 70, 97, 113, 121
Ely 128
Ephesus 33
Erineo 92
Essex 1, 121, 168, 172-180
Exeter 112
Eton 91
Europe 1, 47, 70, 92, 155-157, 159, 167
Fort William (Bengal) 50
Fort York 7
France 64, 70, 73, 76, 131
Fulneck (Moravia) 59
Fulneck (near Leeds) 58-62, 88, 90, 117
Gabon 153
Gaul 68
Geneva 170
Georgia 22, *
Georgian Isles 81
Germany 59, 73, 91, 149, 154, 159, 160, 162,169, 170, 172, 180
Glamorgan 180
Glasgow 113, 173
Gloucester 105
Goa 50
Gold Coast 45
Grace Hill 59
Grand Canaries 9
Great Horkesley 122
Great Houghton 63
Great Ouse, River 20
Greece 85, 86, 159
Greenland 78, 85, 87
Greenwich 40
Guinea 48
Hague, The 149
Hallamshire 92
Hampstead 29, *
Harrogate 86
Hastings 147
Haworth 54
Hereford 122, 140, 145
Herefordshire 122-125, 133, 137, 144, 145
Herrnhüt 102
Hiroshima 161
Hodnet 126
Holborn 127
Holland 70
Hoxton 40, 44
Huddersfield 28
Hudson's Bay 7
Hull 113, 169
Humberside 168, 169
India 50, 88, 89, 104, 126, 154, 160, 170, *
Iona 166
Ireland 7, 59, 124, 144, 178
Irvine 58, 59, 69
Islington 56
Israel 121
Italy 131
Jamaica 3, 8, *
Japan 161, 162
Jericho 116
Jerusalem 170
Kent 3, 4, 7, 8, 13, 28, 53, 101, 126
Kildare 124
Killarney 144
Kingston upon Hull 113, 169
Labrador 78
Lambarene 153
Lancashire 164-172
Lane End (Staffordshire) 83
Laodicea 33
Lebanon 131
Leeds 17, 59, 89, 97
Leigh (Essex) 121
Leipzig 91
Leith 67
Leominster 122, 124
Lichfield 74, 163
Lincoln 140
Linkinhorne 54
Littlemore 122, 128
Liverpool 3, 7-9, 13, 15-19, 28, 47, 74, 92, 93,107, 115, 142, 171, *
London 7, 22, 27, 28, 39-41, 43-45, 52, 54-57, 62, 63, 91, 94, 104, 124, 127-129, 131-134, 139, 143, 157, 164, 169, 171, 174, *
Lough Swilly 7
Lyke-lake 13
Madagascar 105, 166, 170
Made(i)ra 4, 9
Maidstone 3
Malawi 170
Malvern 139
Mana 14
Manchester 65, 74, 89, 176
Margate 90
Martham 22
Matlock Bath 65
Mediterranean Sea 1

Ielton Mowbray 50
Iexico 131
Iiddlesex 101
Iilford Haven 104
Iirfield 61
Ionkland 122-124, 126, 129, 132-137, 140, 144, 145
Ionkseaton 151-156
Iorocco 131
Iorpeth 155-158, 164
Iagasaki 161
Iaples 131
Iewbury 107
Iewcastle-under-Lyme 113
Iewcastle upon Tyne 99, 151, 153-154, 160-161
Iew Delhi 170
Iewfoundland 129, 147
Iewland 139
Iewport Pagnall 27, *
Iew South Wales 92
Iew Zealand 131
Iorthampton 19, 22, 37, *
Iorth Curry 121
Iorthfield 126
Iothumberland 151-162, 164
Iorway 85
Iottingham 123, 173
Iottinghamshire 123, 131, 173
Iyasaland 170
)berammergau 180
)ckbrook 113
)lney xii, 20-25, 27-30, 35, 37, 38, 40, 43, 45, 98, 104,132, *
)rleton 122
)undle 122
)xford 25, 121-122. 126-128, 132, 143, 149, 150, 168, 172
'akistan 178
'apua 153
'eak District 65, 101
'eldon (Essex) 121
'ennine Hills 166
'erth (Scotland) 113
'hiladelphia (Asia Minor) 33
'hiladelphia (USA) 66, 94
'imlico 128, 133, 134, 139
'ixley 128
'lantanes 7
'oland 154
'oldhu 147
'lymouth 4
'ort Sancto 9
'ortugal 45, 70
'ottendorf 59
Ratingen 172
Reading 27, 54, 90
Red Sea 121
Richmond 92
Riviera, The 152
Rochester 8
Rochfort 79
Rodborough 173
Rome 46, 87, 132, 143, *
Romsey 141
Rotherham 81
Russia 43, 70, 159, 170, *
Saffron Walden 172, 175
Saint Albans 9,*
Saint Christopher 44
Saint Helens 169-172, 173
Saint Johns (Newfoundland) 147
Saints Kitts 15, 16, 78
Saint Pancras 143
Salisbury 124, 131, 179
Sandy Point (St. Kitts) 15
Savanna 22
Saxony 91
Scandanavia 72
Scarborough 68, 74
Scotland 50, 58, 69, 90, 93, 107, 115, 166, 173, *
Shebar 11, 15
Sheffield 64-67, 69-72, 74-77, 79-81, 83, 85, 87, 88, 89, 92, 95, 96, 99-102, 104, 107, 109, 115 118, 119, 126
Shoals of St. Anne 12
Shooter's Hill 101
Sicily 131
Sierra Leon(e) 7, 9, 10, 47
Snowdon, Mount 104
Sodom 26
Somerset 121
South Africa 78, 91-92, 95, 104, 105, 113, 149, 170, 171
Southampton 50
South Carolina 8, 46
South Sea Islands 88, 91, 104, 158
Southwell 123, 131
Spain 1, 2, 70, 76, 131
Staffordshire 83
Staithwaite 28
Stepney 16
Stirling 173
Stock 44
Stockholm 149
Stoke Newington 139
Stolberg 91
Stratford on Avon 39, *
Swanland 168, 169
Swilly, Lough 7
Swinton 62
Switzerland 72, 170
Switz, Vale of 72
Syria 131
Tahiti 88
Tenariff 9
Tenbury 136

Teston 53
Tewkesbury 105
Thames, River 101
Thaxted 146, 172-179
Thebes 70
Thriberg 99
Tobago 61, 62, 113
Torbay 4
Trinity Bay (Newfoundland) 129
Tropic of Cancer 13
Tsangchow 153
Turkey 70
Tweed, River 67
Twickenham 74
Tyneside 153, 155
Ulm 73
Ulster 59
United States of America 149, 153
Uppsala 178
Venice 3, *
Vietnam 178
Wales 104, 149, 175, 180
Wansbeck, River 155
Wantage 139
Waterlooville 147, 175
Wath-upon-Dearne 61-63, 66
Wells (Somerset) 143
Wentworth 61
West Hartlepool 154
West Indies 3, 8, 15, 16, 44, 46, 61, 62, 75, 76, 78, 113
Westminster 41, 174
Weston Underwood 29, *
Weymouth 122
Whiston 172
Whitley Bay 151-155
Winchester 122, 128
Windward Coast 47, 48
Worcester 99, 113
York 67, 68, 84, 99, 124
Yorkshire 17, 19, 20, 28, 54, 59, 61, 67, 68, 81, 100, 101, 122, *
Zambia 175
Zinzenorf 59

GENERAL INDEX

An * indicates than an entry also appears in the General Index in *Hymnwriters* I

A (*Use the second word to search index*)
Abdullah or *The Arabian Martyr* (Barham) 90
Aberfan Disaster 175
Aborigines 112
According to Thy gracious word (Montgomery) 99, 100
Accusers Challenged (Newton) 48
Aquittal of the Seven Bishops, The (Hall) 92
Adeste, fideles 132
Advent 131 , 157
Aesop's Fables 11, 12
African, SS 13, 15
African Valley, An (Montgomery) 95
Albert Schweitzer, the Man and his Mind (Seaver) 153
Alexander's Bank 90
All glory, laud and honour (Neale) 128
All Saints, Margaret Street 128, 132
All Souls' College, Oxford 121
Almighty God, whose only Son (Baker) 136
Altho' on massy pillars built (Newton) 28
Amazing grace! (how sweet the sound) (Newton) 6
American Indians 22
American War of Independence 32, 37, *
Amethyst, The 112
Amnesty International 175, 178
Amulet, The (Hall) 92
An (*Use the second word to search index*)
Anaesthesia 28
Angels from the realms of glory (Montgomery) 80, 81
Anglia Television 180
Animals in Heaven 35
Annals of Horsemanship (Carey) 93
Anti Corn-Law League 91
Anti-Semitism 161, 167
Anthem of the Four Living Creatures 121
Apocryphal Chapter on the History of England, An (Montgomery) 96, 97
Apollo II 178
Apologia (Newton) 33, 34, 41
Approach, my soul, the mercy-seat (Newton) 36
Argyle Chapel, Bath 113
Armed Forces 4, 10, 11, 15, 135, 156-158, 161
Armenians 41
Ascension 106
Ask what I shall give thee (Newton) 17, 26
Assemblies of God 164
Astronomy 148, 166
Athanasian Creed 125, *
At Home in Heaven (Montgomery) 114, 115
At the End of the World 152
At the name of Jesus (Noel) 139-141
Authentic Narrative, An (Newton) 1-8
Away with gloom, away with doubt (Shillito)168
Bank of England, 53, 57
Bank Station 57
Baptists 16, 22, 41
Bath Royal Literary and Scientific Institution 113, 119
Battle of Alexandria, The (Montgomery) 70
Battle of Blenheim, The (Southey) 94
Battle of Britain 157
Battle of Leipzig 91
Battle of Sempach 108
Battle of Trafalgar 73
BBC Hymnbook 163
Bear-baiting 35
Beaufort Almshouses 139
Bee, The, SS 16
Begone unbelief (Newton) 14, 36
Be known to us in breaking bread (Montgomery) 97
Bells 124, 129
Benedicite of the Three Children 121
Benediction for a Baby, A (Montgomery) 105
Benefit Societies 99, 132, 135
Berwick Jail 24
Bethesda Chapel, Burnley 164
Bible 1, 14, 25, 29, 32, 43, 50, 78, 86-88, 121,152, 166, 171, *
Bible Associations 78, 90, 105
Birds (Montgomery) 108, 111
Birth Control 176, 177
Birth from Birth (Bayly) 180
Blackburn Academy 93
Black Letter Saints 131
Blessing on our pastor's head, A (Montgomery) 118
Blindness 108, 179, 180
Boers (*See* Boors)
Bohemians 86
Bonfire Night 37
Book of Common Prayer 14, 126
Book of Creation (Newton) 8, 36
Book of nature open lies (Newton) 8, 36
Boors 147
Borrowed Axe, The (Newton) 26
Brasenose College, Oxford 128, *
British and Foreign Bible Society 90, 105, *
British Broadcasting Corporation 163
British Museum 132
Brownlow, SS 7, 8
Buckland Congregational Church 149
Bulgarians 85
Bull Inn, Wicker 64, 76
Burnley Express and News 167, 168

Burns and Plastic Unit, Whiston Hospital, Near St Helens (Bayly) 172
Bury Field 39
Cabinet Cyclopaedia, The (Lardner) 112
Cain and Abel (Newton) 26
Calvinism 35, 41, 179, *
Canary, The (Montgomery) 111
Candid Declaration of the Church known by the Name of the Unitas Fratrum (Spangenberg) 59
Capital Punishment 51
Captains of the saintly band (Baker) 129, 130
Carlisle Cathedral 124
Carver Street National School, Sheffield 118
Castle and Falcon 43
Cat's Boat Club 149
Caustonia: Discourse IX (Bowman) 22
Characteristic Treatise V (Newton) 2, 3
Charles Square, Hoxton 40, 46
Chelsea Parish Church 127
Chemistry 74
Chichester Cathedral 93, *
Child(e), The (Newton) 37
Child Psychology and Modern Sunday School Methods 149
Children recalling Christ's Example and his Love (Montgomery) 117
Chief Shepherd of thy chosen sheep (Newton) 21, 22, 32
Chimney Boys 74, 75, 78
China Evangelised (Montgomery) 115, 116
Cholera 107, 108
Cholera Mount, The (Montgomery) 108
Christ Church College, Oxford 126, 127
Christ Church, Eccleston 171
Christ Church, St. Pancras 143
Christ Church United Reformed Church, Chelmsford 179
Christian Action 171
Christian Keepsake, The (Ellis) 91
Christian Oratory, The (Bennet) 2
Christian Psalmist, The (Montgomery) 88, 97, 98
Christian Psalmody (Bickersteth) 93, 107
Christian Research in Asia (Buchanan) 50
Christian Sisterhood, The (Montgomery) 117
Christian World, The 166
Christ in Thaxted (Bayly) 174
Christmas Box, The 81
Christ our Example in Suffering (Montgomery) 88
Chronicle of Angels, The (Montgomery) 109, 111
Church Hymnal (Denton) 127
Church Hymn and Tune Book (Blew) 127
Church Hymns (Ellerton et al.) 107, 108, 140, *
Church Militant learning the Church Triumphant's Song, The (Montgomery) 96
Church Missionary Society 90, 94, 168
Church of England 19, 33, 34, 40, 41, 54, 79, 85, 121-123, 125-127, 130-132, 143, 144, 154
Church of England Psalm-Book, A (Kennedy) 92
Church Training Corps 161
Cinderella 154
Civil War 70, 86, 109
Climbing Boys' Soliloquies, The (Montgomery) 74, 75
Clothing Clubs 140
Coals for the Poor 140
Coleman Street Buildings 46
College for American Indians 22
College of Fort-William 50
Come, my soul, thy suit prepare (Newton) 17
Command thy blessing from above (Montgomery) 80, 81
Commentary on the Psalms (Horne) 112
Common Lot, A (Montgomery) 74
Commons and Footpath Preservation Society 169
Communism 171
Complaint, The (Montgomery) 75
Concise Account of the Present State of the Mission of the United Brethren 78
Confidence (Newton) 37
Confucianism 150
Congregational Council for World Mission 174
Congregational Hymn Book 93, 105
Congregationalists 81, 93, 105, 149, 151, 152, 155, 161, 164, 174
Congregational Union 93, 172
Construction of the World through Faith 150
Convict's Appeal (Barton) 90
Co-operative Societies 151, 152
Copyright 133, 136, 140, 141, 143
Corn Laws 91
Cornwall, SS 9
Cotton 72, 92, 164
Council for the Preservation of Rural England 169
Cox and Box (Burnand/Sullivan) 150
Crucifixion (Stainer) 136
Cry from South Africa, A (Montgomery) 108, 113
Cuban missile crisis 172
Cullercoats Fishermen's Mission Male Voice Choir 153
Cumbrae Collegiate Church 139
Cutlers' Feast 107
Cutlers' Hall 76, 85, 118
Dacre Street Congregational Church, Morpeth 155-162, 164
Daisy in India, The (Montgomery) 88, 89
Danger etc. in the Voyage from Cape Lopez (Newton) 6
Danger of Riches, The (Richards) 133
Daniel (Newton) 26
Day in the Lord's Courts, A (Montgomery) 77
Day thou gavest Lord is ended, The (Ellerton) 107
Death 1, 16, 35, 50, 52, 60, 62, 73, 113, 132, 166,180, *
Death and War, 1778 (Newton) 32, 35
Departed Days (Montgomery) 58, 62

*Derbyshire Tourist's Guide and Travelling Companion, The (*Rhodes) 65
Descent of the Spirit, The (Montgomery) 83, 84
Dial, The (Montgomery) 71, 72
Dialogue on the Alphabet, A (Montgomery) 93, 96, 97
Diamond Classics, The (Pickering) 93
Dibbling 107
Disagreeable Surprise, The (Daniel) 93
Dissenters 1, 16, 19, 28, *
Divine Compassion, The (Bayly/Lloyd Webber) 169
Doncaster Sessions 66
Double Trinity (Bayly) 180
Dr. Barnardo's Homes 174
Dream of the English Church in the Twenty-First Century (Bayly) 152, 180, 181
Druzes 131
Duke of Argyle, SS 8-12
Durham and Northumberland Congregational Union 164
Dykes Memorial Fund 144
Easter Celebrations 60, 61, 139, *
Eastern Orthodox Church 85, 127
East India Company 50, *
Eccleston Congregational (United Reformed) Church 169-172
Ecletic Society 43
Ecumenical Conference of Protestant Churches 149
Ecumenism 149, 166, 171, 178
Eden Cafe, Whitley Bay 151
Edinburgh Monthly Magazine 92
Edinburgh Witness 115
Effort—in another Measure, The (Newton) 36, 37
Elegy in a Country Churchyard (Gray) 44
Elijah fed by Ravens (Newton) 26
Elijah's example declares (Newton) 26
Encyclopaedia Metropolitania (Smedley) 94
Enterprise Neptune 176
Ephesus (Newton) 33
Esau (Newton) 26
Essay on the Universe, An (Browne) 20
Essays on Song Writing (Aikin) 90
Evangelical Magazine 81, 83, 115
Evening Song for the Sabbath-Day (Montgomery) 118
Every-day Tale, An (Montgomery) 108, 110
Exeter College, Oxford 132
Exhortation to Praise and Thanksgiving (Montgomery) 96
Experimental Chemistry (Stancliffe) 74
Extent of Messiah's Spiritual Kingdom, The (Newton) *48*
Eye Ward, Chelmsford and Essex Hospital (Bayly) 180
Fairway Hall, Monkseaton 151-156
Faith's Review and Expectation (Newton) 6
Family Altar, The (Montgomery) 97, 99
Family-Instructor 2
Family Table, The (Montgomery) 97
Father, Thy life-creating love (Bayly) 173
Father, whose great love enfolds (Bayly) 174
Feed the Minds 174
Fellowship of Reconciliation 157
*Field Flower: on finding One in full Bloom on Christmas Day, A (*Montgomery) 73
Field of the World, The (Montgomery) 106
Fifty-Fifty Girl, The 152
Food and Agricultural Organisation 171
Food and Hunger 156-162, 171, 175, *
Food, raiment, dwelling, health and friends (Montgomery) 97
For a female Friendly Society (Montgomery) 99, 100
For all the saints who from their labours rest (How) 128
For a Meeting of Ministers (Montgomery) 108
For a Solemn Assembly (Montgomery) 80, 81
For Ascension Day (Montgomery) 106
Force of Tomorrow, The (Bayly) 168
Force of Truth, The (Scott) 29, *
Ford Motor Company 168
For Ever (Bird) 91
For ever with the Lord (Montgomery) 112, 114, 115
Forget-me-not (Ackermann) 90, 91
Forster Collection, The 91
For the Peace and Prosperity of the Church (Montgomery)112
Free Church Choir Union 174
Friendly Societies 85, 99, 135
Friend that sticketh closer than a Brother, A (Newton) 31
Frogmore Street Dispensary 91
From glory to glory advancing, we praise thee, O Lord (St. James/Holst) 173
Fulneck Moravian Settlement 58, 59, 62, 88, 90, 118
Gabriel Silvertongue 64, 68
Galvanism 74
Gateacre Comprehensive School, Liverpool 171
Gates of the Paradisical Garden of Roses, The (Boeheme) 32
Geistreiches Gesangbuch 129
Gentleman's Magazine 20
Gethsemane (Bayly) 174
Gideon's Fleece (Newton) 26
Girls' Friendly Society 153
Glad was my heart to hear (Montgomery) 109, 112
Glass Manufacture 169
Gloria Patri 37, 125
Glorious things of thee are spoken (Newton) 30
Glow-worm, The (Montgomery) 73
God is my strong salvation (Montgomery) 88, 89

God moves in a mysterious way (Cowper) 36
God of your forefathers praise, The (Montgomery) 102, 104
God's Universe (Bayly) 179
God the Father, from Thy throne (Baker) 132
Golden Calf, The (Newton) 26
Good Man's Monument, A (Montgomery) 80
Good Tidings of Great Joy to All People (Montgomery) 80
Go to dark Gethsemane (Montgomery) 88
Goodwill Day 171
Gospel Magazine 42
Grace Hill Moravian Settlement 59
Grave, The (Blair) 59, 60, 73
Grave, The (Montgomery) 73
Great Advent, The (Newton) 52
Great Fire of London 39
Great House, The, Olney 32
Great Lord of the Universe (Bayly) 178
Great Shepherd of thy chosen sheep (Newton) 21, 22, 32
Great Shepherd, The (Newton) 47
Great Western Railway 127
Greek Language 19, 25, 51, 127
Greenland (Montgomery) 85-87
Greyhound, SS 5
Guardian, The 128, *
Guide Movement 154, 160
Guild of Church Workers 140
Guild of Holy Living 140
Guy Fawkes Night 37
Guy's Hospital, London 28
Gypsies 112
Gypsy Moth IV 176
Haddo House 124
Hail to the Lord's Anointed (Montgomery) 88
Hallamshire: the History and Topography of Sheffield (Hunter) 92
Hallelujah! (Montgomery) 83
Hannah (Montgomery) 63
Hannah (Newton) 63
Hark! how time's wide sounding bell (Newton) 32, 35
Hark, my soul! It is the Lord (Cowper) 96
Hark! the song of Jubilee (Montgomery) 83
Hartshead, Sheffield 66
Harvest Thanksgiving 135, 156, 158, 159, 161
Harwich, HMS 4, 16
Hastings Grammar School 147
Hear what the Lord, the great Amen (Newton) 33
Heaven 35, 61, 107, *
Heaven in Prospect (Montgomery) 107
Hebrew Language 19, 25
Hereford Infirmary 140
Hereford Journal 122, 145
Herefordshire Church Union 140
Heron, The (Montgomery) 111
He walks unseen (Bayly/Lloyd Webber) 169
Hindoos 102, *
Hindoostanee Language 50
Hints concerning Title-Pages, and A Whisper about Whispering (Montgomery) 68
Historic Gallery, Pall Mall, The 76
History of a Church and a Warming-Pan, The (Montgomery) 64
History of Christian Missions from the Reformation to the Present Time (Huie) 113, 115
History of the Rebellion (Clarendon)109
HMS Harwich 4, 16
HMS Surprise 10, 11
Hollingreave (Bayly) 164, 165
Hollingreave Congregational Church, Burnley, 165, 166
Holy, Holy, Holy Lord (Montgomery) 105
Horkesley House, Monkland 123, 124, 133, 136
Hottentots 95
How hurtful was the choice of Lot (Newton) 26
How sweet the name of Jesus sounds (Newton) 31
How welcome was the call (Baker) 132, 133
Huguenots 93, *
Human Rights 175
Hunger (See Food and Hunger)
Hymnal for Use in the English Church, A (Murray) 123, 126
Hymnal Noted (Neale) 127, 129
Hymn for Amsterdam, A (Bayly) 166, 167
Hymn for Benefit Societies, A (Baker) 135
Hymn for Homemakers, A (Bayly) 165
Hymn for Prisoners of Conscience, A (Bayly) 178
Hymn for the Opening of Sheffield General Infirmary (Montgomery) 70
Hymn for the Wesleyan Centenary, A (Montgomery) 117
Hymn of Zacharias 121
Hymns Ancient and Modern 121-144, *
Hymns and Introits (White) 127, 139
Hymns and Songs of the Church (Wither) 129
Hymns fitted to the Order of Common Prayer (Pott) 129
Hymns for Mission Churches (Baker) 144
Hymns for the London Mission (Baker) 142
Hymns for the Sundays and Holy Days of the Church of England (Woodford) 128
Hymns from the Parisian Breviary (Williams) 127, *
Hymns from the Roman Breviary for Domestic Use (Mant) 127
Hymns of the Church, mostly Primitive (Chandler) 127
Hymns, partly collected and partly original (Collyer) 78
Hymn Society of Great Britain and Ireland 180, 181
Hymns, written and adapted to the Weekly Church Services of the Year (Heber) 126

f Christ were born in Burnley (Bayly) 165, 166
f Solomon for wisdom pray'd (Newton) 26
f the Lord our leader be (Newton) 26
ncome Tax 80
ndependents 16, 27, 28, 33, 34, 41, 155
nferno (Dante) 108
n Memory of Albert Bayly (Green) 181
nner Vision, The (Bayly) 180
n Old Kentucky 152
nquisition 50
nter-Church Aid and Refugee Service 171
n the hour of trial (Montgomery) 107, 108
n the power of God's own Spirit (Bayly) 179
ntroductory Essay on 'The Pilgrim's Progress' (Montgomery) 115
ona Abbey 166
ris, The 66-68, 70-81, 85-87, 91, 92, 99, 102
rvine Moravian Church 59
will trust and not be afraid (Newton) 14
acob's Ladder (Newton) 26
ames Montgomery (King) 65
esus, grant me this, I pray (Baker) 129
esu, lover of my soul (Wesley) 122
esus Christ, I love and serve (Bayly) 180
esus our best-beloved Friend (Montgomery) 78
esus, where'er thy people meet (Cowper) 32
ew, The (Cumberland)70
ohn Williams V 158
onathan Starlight 64, 65
oseph made known to his Brethren (Newton) 5
ourney to Kent (Newton) 7
ugurthine War 46
uvenile Delinquency (Montgomery) 96
Kendall Gale (Bayly) 166
Key to the Universe (Calder) 179
Kindly) Spring again is here (Newton) 32, 34
Kindred in Christ, for his dear sake (Newton) 32, 35
King of Love my Shepherd is, The (Baker) 136, 142, 144
Kingsway Hall, Bow 154
Lace-Making 20, *
Lady Huntingdon's Chapel 41
Lamb, SS 9
Lambarene Hospital 153
Lancasterian Schools, 97, 118, *
Langbourn-Ward Charity School 54, 55
Laodicea (Newton) 33
Latin Language 1, 19, 43, 127, 129, 149
Law Courts, London 124
Law read by Ezra and the Covenant Renewed, The (Montgomery) 95
Lay Preachers' Training 180
League of Nations 154, 174
League of Nations Union 154, 155
Leisure 152
Le Mans Breviary 129
Leominster Town Hall 122
Leon, SS 108
Le Pauvre Jacques (Ryan) 90
Leprosy 179
Le Rodeur, SS 108
Let songs of praise arise (Montgomery) 104
Letters to a Nobleman (Newton) 22
Let us love, and sing, and wonder (Newton) 37, 38
Let us with a gladsome mind (Milton) 132
Lichfield Cathedral 74, 163
Life is born of death to-day (Bayly) 180
Life of a Flower, The (Montgomery) 96, 97
Life of Charles Wesley, The (Jackson) 92
Lift up your heads, ye gates of brass (Montgomery) 115, 116
Light of the World (Sullivan) 142
Light shining out of Darkness (Cowper) 36
Lilac Time 168
Lima Choir 173
Lion that on Samson roar'd, The (Newton) 26
Litany of Intercession (Baker) 131, 132
Liturgy and Hymns for the Use of the Protestant Church of the United Brethren 115
Liturgy of Saint James 173
Liverpool Academy of Arts 74, 92, 93
Liverpool Festival 142
Liverpool Scotch Session Church 107
Lives of Individuals who raised Themselves from poverty to Eminence and Fortune (Davenport) 92
Lock Chapel 41
London College of Music 169
London Missionary Society 81, 88, 91, 158, 162, 171, 174
London Society for the Blind 136
London Underground 56, 57
Look on the Fields (Bayly/Lloyd Webber) 169
Lord, are these eyes that see the sun (Montgomery) 78, 79
Lord Falkland's Dream (Montgomery) 107-109
Lord, for ever at Thy side (Montgomery) 117
Lord, from whom beauty, truth and goodness spring (Bayly) 146
Lord God, the Holy Ghost (Montgomery) 83, 84
Lord Jesus, God and Man (Baker) 124, 125
Lord of the boundless curves of space (Bayly/Wren) 148
Lord of the home, Thine only Son (Bayly) 165
Lord, pour Thy Spirit from on high (Montgomery) 107
Lord, save Thy world in bitter need (Bayly) 166
Lord, teach us how to pray aright (Montgomery) 82, 83
Lord, Thy Kingdom bring triumphant (Bayly) 167
Lord, thy word abideth (Baker) 132
Lord will provide, The (Newton) 8
Lot in Sodom (Newton) 26
Lucan Characters 136
Lutherans 59

Lyre, The (Montgomery) 71, 73
Magnificat 121
Mahratta Language 50
Malayan Language 50
Manchester Exchange 89
Manchester Philological Society 74
Mansfield College Magazine 149, 150, 168, 174
Mansfield College, Oxford 149, 151, 152
Manuscript Magazines 78
Maoris 131
Margaret Street Chapel 128, 132
Margate Sea Bathing Infirmary 90
Mariners' Compass 9
Marmion (Scott) 108
Maronite Christians 131
Marriage 132
Martha and Mary (Newton) 33
Martha her love and joy express'd (Newton) 33
Matrimonial Holiday 152
Mayor of Donchester, The (Montgomery) 64, 65
May the grace of Christ our Saviour (Newton) 37, 39
Memnon's Statue 70
Memoirs of the late Mrs S. Huntingdon of Boston (Wisner) 104, 105
Memoirs of the Public and Private Life of John Howard, the Philanthropist (Brown) 92
Messiah (Handel) 47-49
Messiah: Fifty Expository Discourses (Newton) 47-49
Methodists 17, 25, 29, 40, 41, 43, 50, 59, 88, 99, 151, 161, *
Metrical History of Manchester, A (Aston) 89
Metropolitan Magazine 60, 61
Micah (Bayly) 166
Millions within Thy courts have met (Montgomery) 118
Mingled Character of the Divine Dispensations, The (Raffles) 107
Minstrelsey of the Scottish Border (Ballantyne) 90
Missionaries 59, 61, 62, 78, 81, 85, 87, 88, 90, 92, 113, 118, 135, 166-168
Missionary Register (Pratt) 90
Modes of Government (Montgomery) 64
Money Markets 92
Monkland Church 120, 133, 134, 136, 137, 144, 145
Monkland Parish Magazine 144
Monkland School 124, 126, 140, 144
Monkland Vicarage 123, 124, 133, 136
Monkland Village Hall 124
Montgomery Friendly Society 99
Montgomery Silver Medal 118
Montgomery Memorial 89, 119
Moon in silver glory shone, The (Newton) 28, 29
Moralists, a Philosophical Rhapsody, A (Cooper) 2, 3
Moravian Hymn Tunes 113
Moravians 19, 27, 41, 59-61, 78, 79, 85-88, 113, 115, 118
Morden College, Blackheath 20
Morpeth Community Council of Social Service 162
Morpeth Full Gospel Church 164
Morpeth Herald and Reporter 157-162
Morpeth Ministers' Club 160
Morpeth Peace Group 157
Most high and holy Lord (Bayly) 174
Mount, The, Sheffield 102, 115
Mozley's Hymnal 123, 126
Musical Meditations (Barham) 90
Music in Thaxted (Bayly) 174
Music of the Church considered in its various Branches Congregational and Choral (La Trobe) 90
My barns are full, my stores increase (Newton) 33
My Father, for another night (Baker) 142
My God, I love Thee not because (Xavier/Caswell) 138
*Mysterious Universe (*Atkins) 153
Nahum (Bayly) 166
Name of Jesus, The (Newton) 31
Name of Jesus, and other Verses for the Sick and Lonely, The (Noel) 139
National Health Service 179
National Schools 97, *
National Trust 176
Nazis 154, 162
Negro is free (Montgomery) 113
Negro's Complaint, The (Cowper) 48, 53
New Bayley Prison, Manchester 89
Newcastle Courier 99
New College, Oxford 122
New every morning is the love (Keble) 129
New Monthly Magazine (Shoberl) 90
Newton watches over him (Newton) 24
New Version (Brady/Tate) 125
Norfolk Street Upper Chapel, Sheffield 66
North American 112
Northern Line, London Underground 57
Norton Bowling Green, Sheffield 76
Nottingham Ecumenical Conference 173
Nottinghamshire Church Choral Union 123
Novello & Co. 136
Now, my soul, thy voice upraising (Baker) 129
Nunc Dimittis 121
Oberammergau 1980 (Bayly) 180
Observations on Friendly Societies (Cunningham) 94
Occurrences in Early Life (Newton) 1-4
O Christ, Redeemer of our race (Baker) 129
O come all ye faithful (Oakeley) 132
Ode to Apathy (Hoole) 75
Ode to the Volunteers of Britain on the Prospect of Invasion (Montgomery) 73
Of Meditation (Bennet) 2

Of the practical Influence of Faith (Newton) 33
O God of love, O King of peace (Baker) 131, 132
O God, Thou art my God alone (Montgomery) 88, 95
O God, whose love in anger burns (Bayly) 178
O Holy Ghost, Thy people bless (Baker) 143
Oh! what if we are Christ's (Baker) 123
Old Church of the Brethren 59
Old Church Psalmody (Havergal) 127
Old Version (Sternhold and Hopkins) 125
Old Woman (Montgomery) 96
Oliver's Cafe, Morpeth 164
Olney Hymns (Cowper and Newton) 24, 25, 29, 40, 53, 98, 104, 132, *
Olney Hymns, The (Newton) 32
Olney Independent Church 28
Olney Parish Church xii, 21
O Lord of circling planets and all the stars in space (Bayly) 176
O Lord of every shining constellation (Bayly) 166
Omicron 42
On Covetousness (Newton) 51
One Hundred Years of Congregationalism in Morpeth 156
One song of praise, one voice of prayer (Montgomery) 113, 117
One there is, above all others (Newton) 31, 56
On Female Dress (Newton) 51
On His pilgrimage of woe (Montgomery) 117
On the Earthquake, September 8, 1775 (Newton) 28
On the Eclipse of the Moon, July 30, 1776 (Newton) 28, 29
On the Fire at Olney, September 22, 1777 (Newton) 30
On the Phrenology of the Hindoos and Negroes (Montgomery) 102
On the Propriety of a Ministerial Address to the Unconverted (Newton) 49
On the Snares and Difficulties attending the Ministry of the Gospel (Newton) 38
On this day, the first of days (Baker) 129
On Trust in the Providence of God, and Benevolence to his Poor (Newton) 28
O praise our God to-day (Baker) 132
Orchard Side, Olney 24
Oriel College, Oxford 128
Orissa Language 50
Original Hymns for Christian Worship (Montgomery) 77, 81, 82, 84, 88, 94, 95, 97, 100, 105, 106, 118
O sacred head, surrounded (Baker) 129
O Spirit of the living God (Montgomery) 94, 95
O thou by whom we come to God (Montgomery) 83
Ottoman Empire 87
Our Saviour's Miracles: Sketches in Verse (Montgomery)113
Our souls shall magnify the Lord (Montgomery) 99, 100
Out of the deep I call (Baker) 136
Oxford Diploma of Theology 150
Oxford Movement (*See* Tractarian Movement)
Oxford University 121, 122, 126-128, 143, 151, *
Pacifism 147-149, 152, 154, 157, 160
Palms of Glory, raiment bright (Montgomery) 104, 107
Papal Infallibility 177
Paris 1815 (Croly) 94
Paris Breviary 127
Park Avenue Congregational Church, Whitley Bay 151-155
Parliament 76, 80, 93, 94, 99, 112, *
Parliamentary Committee on Aborigines 112
Parochial Psalmody (Goss) 127
Patriotism 153
Patriotic Song by a Clergyman of Belfast 66, 67
Patriot's Pass-Word, The (Montgomery) 108,110
Paul Positive 68
Peace and War 32, 68, 70, 73, 76, 79, 81, 91, 147, 149, 152, 154, 157, 160
Peace in Thaxted (Bayly) 177
Peeps from the Past (Ward) 75
Pelican Island, The (Montgomery) 102, 103, 113
Pembroke Place, Liverpool 115
Perseverance (Newton) 37, 39
Persian Language 50
Pews 127
Philadelphia (Newton) 33
Phrenology 102
Pilgrim Fathers 179
Pilgrim's Progress, A (Bunyan) 24, 54, 101, 115
Pilots, The 158, 171, 172
Plan of Academical Preparation for the Ministry, A (Newton) 18, 19, 42
Plan of a compendious Christian Library, A (Newton) 42
Pleasing Spring is here again (Newton) 32, 34
Pleasures of Imprisonment, The (Montgomery) 68
Poems of William Cowper, Esq., Introduction to (Montgomery) 97
Poetry and Poets (Ryan) 90
Poet's Portfolio, A (Montgomery) 107-113, 115
Polish Flood Relief Fund 154
Poor Esau repented too late (Newton) 26
Pope's Encyclical 177
Pope's Willow (Montgomery) 74
Portfolio, The 79
Pour out Thy spirit from on high(Montgomery) 107, 108
Praise and Thanksgiving (Bayly) 173
Praise for redeeming Love (Newton) 38
Praise O praise our God and King (Baker) 132
Praise the Lord through every nation (Montgomery) 106
Prayer for an aged Minister (Montgomery) 118

Prayer for Humility (Montgomery) 117
Prayer for Ministers (Newton) 21, 22
Prayer is the soul's sincere desire (Montgomery) 82, 83
Prayers on Pilgrimage (Montgomery) 108
Pre-Eminence of Poetry, The (Montgomery) 104
Preparation of the Heart, The (Montgomery) 82
Prepare to meet God (Newton) 15, 36
Presbyterians 19, 41, 149, 161, 179
Prince Arthur (Blackmore) 60
Principles of Christianity as taught in Scripture, The (Bowman) 22
Prison Amusements (Montgomery) 68
Proclaim the year of Jubilee (Montgomery) 115, 117
Prolegomena (Kennicott) 42
Prophets' sons, in time of old, The (Newton) 26
Prose by a Poet (Montgomery) 93, 95, 97
Psalm 23 136
Psalms and Hymns (Watts) 98
Psalm Tunes for Cathedrals and Parish Churches (Crotch) 127
Psalter pointed and set to accompanying Chants 'Ancient and Modern' (Monk/Baker) 137
Psychology and the subjective View of Religion 150
Public Dinner for James Montgomery 99, 101, 102
Quakers *(See* Society of Friends)
Questions and Answers (Montgomery) 58, 61, 62, 87, 90, 118
Quiet, Lord, my froward heart (Newton) 36, 37
Race Relations 149, 167
Reading Mercury 90
Red Cross 157-160
Redeem'd restored, forgiven (Baker) 145
Redshirts 131
Refugees 72, 104, 161, 171
Reign of Christ on Earth, The (Montgomery) 88
Rejoice, believer in the Lord (Newton) 37, 39
Rejoice, O People (Bayly) 162, 163
Rejoice, O people, in the mounting years (Bayly) 162, 163
Religious Doubt (Bennett) 149, 150
Religious Drama 155
Religious Tract Society 81, 115
Remembrance and Resolution (Montgomery) 95
Review of Ecclesiastical History, A (Newton) 10, 22, 25, 28
Reynolds Commemoration Society 81
Richmond College 92
Rights of Man (Paine) 64
Riviera, The (Wills) 152
Road Safety 169, 170
Robert Burns (Montgomery) 69
Rock of ages, cleft for me (Toplady) 35, 122, *
Rodborough Hymnbook 173
Rodborough Tabernacle 173
Roman Breviary 127
Romanism 33, 40, 50, 61, 122, 132, 143, 177, *
Rose Cottage, Monkseaton 151
Rotherham Independent College 81
Royal Academy 93
Royal Berkshire Hospital 54
Royal Dockyard, Portsmouth 149
Royal Dockyard School 147
Royal Institution 104
Royal Navy 4, 10, 11, 15, 135
Russian Orthodox Church 170
Sacred Melodies from Haydn, Mozart, Beethoven, and Other Composers (Gardiner) 76, 91
Saffron Walden and District Free Church Federal Council 175
Saffron Walden Friends' School 172
Saint Andrew, Holborn 127
Saint Barnabas Clergy House, Pimlico 128, 133
Saint Barnabas Hostel, Newland, 139
Saint Barnabas, Pimlico 128, 133, 134, 139
Saint Brigid, Kildare 124
Saint Cecilia's Day 164
Saint Cuthbert's Presbyterian Church, Whitley Bay 155
Saint Dionis-Backchurch, London 131, 143
Saint Edmund's House, Oxford 25
Saint Giles, Reading 27
Saint James' Congregational Church, Newcastle 153
Saint James' Day 123
Saint James, Manchester 89
Saint James, Morpeth 161
Saint John Ambulance Brigade 158
Saint Mary Hall, Oxford 143
Saint Mary, Nottingham 123
Saint Mary Woolchurch-Haw, London 40, 55
Saint Mary Woolnoth of the Nativity, London 39-41, 44, 50-57
Saint Michael's College, Tenbury 136
Saint Paul's Cathedral, London 52, 136, 140, 164, *
Saint Paul, Sheffield 83, 126
Saint Stephen, Walbrook 94
Salem Chapel, Burnley 164
Salisbury Cathedral 124, 179
Salisbury Diocese 128, 131, 134
Samaritans, The 94
Samson's Lion (Newton) 26
Sarum Breviary 129
Sarum Hymnbook 131
Science 74, 148, 166, 171, *
Sclavonic Tribes 85
Scorched Earth 147
Scotch Session Church 107
Scouting 153, 160
Sea Bathing Infirmary 90
Sea Piece, A (Montgomery) 96, 98
Second Epistle to a Friend (Montgomery) 68
Selection of Psalms and Hymns, A (Cotterill) 83-85, 126

elect Scottish Songs, Ancient and Modern (Burns/Cromek) 92
elect Works of John Bunyan 115
hall we not love thee, Mother dear (Baker) 135, 136
heffield Aged Females Society 118
heffield Book Club 76
heffield Boys' Charity School 87, 118
heffield Cathedral 118
heffield Club 115
heffield Courant 67
heffield Female Friendly Society 85
heffield General Infirmary 69-71, 79, 80
heffield Girls' Charity School 87, 118
heffield Independent 99
heffield Iris 99. *(See also 'Iris, The')*
heffield Lancasterian Schools 118
heffield Mechanics' Institute 115
heffield Mercury 74, 99
heffield Michaelmas Sessions 67
heffield Park 109
heffield Philosophical and Literary Society 89, 96, 102, 104
heffield Register 63-66
heffield School for Industry 71
heffield School of Anatomy and Medicine 102
heffield School of Art and Design 118
heffield Society for Bettering the Conditions of the Poor 81
heffield Sunday School Union 81, 104
heffield Theatre 70
he was a city proudly strong (Bayly) 166
horthand 115
ickness and Health 16, 28, 50, 52, 54, 55, 69, 73, 80, 91, 97, 102, 132, 150, 175, 179, 180
igns which God to Gideon gave, The (Newton) 26
ing, Hallelujah, sing (Montgomery) 105-107
ing we the song of those who stand (Montgomery) 96
inner, art thou still secure? (Newton) 15, 36
ion College 143
isters of the Poor 132
ketch of the Life of Rev. John Cowper (Cowper) 56
lavery 4, 5, 8, 10-13, 15, 44-48, 51, 59, 72, 76, 94, 104, 108, 109, *
lavery (More) 51
leeping Children (Chantrey) 74
luggard, The (Newton) 37
ocial Dedication to God (Montgomery) 78
ociety for Promoting Christian Knowledge 52, 140, *
ociety for the Promoting of Useful Knowledge 74
ociety for the Bettering of the Poor in Sheffield 118
ociety for the Education of Young Men for the Work of the Ministry amongst Protestant Dissenters 67
Society for the Propagation of the Gospel 140,*
Society instituted for the Relief of poor, pious Clergymen of the Established Church in the Country 54
Society of All Saints 132
Society of Friends 68, 73, 81, 94, 154, 161
Society of Friends of Literature 66
Society of Saint Timothy 140
Society of the Holy Cross 121, 143
Soldier No Christian, A 73
Sometimes a light surprises (Cowper) 36, 53
Song of Bethlehem (Bayly/Lloyd Webber) 169
Song of Miriam 121
Song of Moses 121
Song of the Lamb 121
Songs of praise the angels sang (Montgomery) 83, 84
Songs of Zion (Montgomery) 88
Songs on the Abolition of Negro Slavery in the British Colonies (Montgomery) 109, 113
Sound the loud timbrel o'er Egypt's dark sea (Moore) 113
South African Commercial Advertiser 92
South African Journal 92
South London Underground Railway 56
Southwell Minster (Cathedral) 123, 131
Sow in the morn thy seed (Montgomery) 106,107
Space Travel 178
Sparrow, The (Montgomery) 111
Spinning 165
Spirit accompanying the Word of God, The (Montgomery) 95
Spiritual Authority—A Chinese Approach to the Question (Hughes) 150
Spring II (Newton) 32, 34
Spring Street Sunday School 77
SS African 13, 15
SS Brownlow 7, 8
SS Cornwall 9
SS Duke of Argyle 8-12
SS Greyhound 5
SS Lamb 9
SS The Bee 16
Stand up and bless the Lord (Montgomery) 95, 96
Stanzas to the Memory of Rev. Thomas Spencer (Montgomery) 76, 77
State Lottery 8, 81
State Lottery, The (Roberts) 81
Stepney Independent Church 16
Steps, Thaxted, The (Bayly) 173
Stockholm Conference on Life and Work 149
Strictures on Mr. Montgomery's Essay on the Phrenology of the Hindoos and Negroes (Thompson)102
Student Prince, The 152
Sunday Schools 21, 77, 83, 97, 104, 151, 154
Sunday Thoughts (Browne) 20
Sun-Flower, The (Montgomery) 108

Supported by the word (Newton) 26
Supposed Meeting of Cowper and Newton in Heaven, The (Newton) 56
Surprise, HMS 10, 11
Swanland (Bayly) 169
Swanland Congregational Church 168, 169
Symphonia Sirenum 129
Syriac Language 19
Tabernacle, The 23, 41
Table Talk 103
Table Talk (Cowper) 97, *
Tale without a Name, A (Montgomery) 108
Tanning 155
Teach me, O Father, faith that leaps (Bayly) 180
Television 180
Temple of Concord 39
Ten Commandments 125
Ten thousand times ten thousand (Alford) 121
Testimony (Bayly) 180
Thaxted Baptist Church 173
Thaxted Congregational (United Reformed) Church 174-178
Thaxted Old Independent Meeting 178
The (*Use the second word to search index*)
Thelyphtor (Madan) 43
There is a book, who runs may read (Keble) 132
There is a garden whither Jesus came (Bayly) 174
Thesbian Society 70
Tho 'troubles assail (Newton) 8
Thoughts preparative or persuasive to Private Devotions (Sheppard) 90
Thoughts upon the African Slave Trade (Newton) 44-48
Three New Carols (1825) 81
Thrice Holy (Montgomery) 106
Thunderstorm, The (Montgomery) 74, 75
Thus saith the holy One, and true (Newton) 33
Thus saith the Lord to Ephesus (Newton) 33
Thy greatness is like mountains, Lord (Bayly) 173
Tin Tabernacle, Whitley Bay 151
'Tis past—the dreadful stormy night (Newton) 37
To Albert Schweitzer (Bayly) 153
Token of Affection and Respect, A (Newton) 40
To my Friend, George Bennett, Esq. (Montgomery) 89
Tontine Inn, Sheffield 99
To Thy Temple I repair (Montgomery) 77, 78
Tottenham Court Chapel 23, 41
Tractarian Movement 122, 124, 127, 132, *
Treatise on Prayer (Bickersteth) 83
Trials 66, 67, 71, *
Trinity College, Cambridge 123, 139
Trinity United Reformed Church, Whitley Bay 155
Triumph of Christianity, The (Abbott) 92
Trust in the Lord (Montgomery) 89
Twelve Articles of Christian Faith 125
Twelve Dramatic Sketches founded on the Pastoral Poetry of Scotland (Hetherington) 93
Ugolino and Ruggieri (Montgomery) 108
Unitarianism 66, 79, 91, *
Unitas Fratrum (*See* United Brethren)
United Brethren 19, 27, 41, 59-61, 78, 79, 85-88, 90, 113, 115, 118
United Methodist Free Church 93
United Nations 171, 174, 175
United Reformed Church 179
United Reformed Church of Canada 149
Universal Chorus, The (Newton) 49
Upper Chapel, Norfolk Street, Sheffield 66
Utile Dulci Club 76
Veneration of the Blessed Virgin Mary 135, 136
Veneration of the Cross 129
Veni Creator 125
Venite 125
Verses on Finding the Feathers of a Linnet (Montgomery) 69, 70
Verses to a Robin Redbreast (Montgomery) 67
Verses to my Friend, George Bennett, Esq. (Montgomery) 88, 89
Victorine's Excursions (Le Noir) 90
Vigil of Saint Mark, The (Montgomery) 69, 70
Vindication of War (Cobbett) 76
Visions of Patmos, The (Grinfield) 91
Voice from the Sanctuary, A (Illingworth) 115
Voice that breathed o'er Eden (Keble) 135, 136
Voyage from Cape Lopez to England (Newton) 5
Voyage of the Blind, The (Montgomery) 108, 112
Voyages and Travels (Montgomery) 104, 105
Voyage to Africa (Newton) 8
Voyage to Madeira (Newton) 4
Voyage to Terra Australis (Flinders) 102
Wall Street Collapse 153
Wanderer of Switzerland, The (Montgomery) 72
War and Peace 32, 35, 46, 66, 68, 70, 73, 76, 79, 81, 91, 109, 131, 147, 149, 152, 155, 157-161, *
Warner & Sons 129
Waterlooville Baptist Church 175
Way to Christ, The (Boheme) 32
Wearied by day with toils and cares (Newton) 30
Weaving 165
We count you now among the pioneers (Green) 181
Welcome to Christian Friends, A (Newton) 32, 35
Wells Theological College 143
We look into the atom and find a world within (Bayly) 179
We love the place, O God (Baker) 129, 130
We love the place, O God (Bullock) 129
West Indies, The (Montgomery) 75, 76
Westminster Choir College 174, 175
What does the Lord require? (Bayly) 166
What every Woman knows (Barrie) 152
What is Christ doing now? (Bayly) 167
What is Prayer? (Montgomery) 82
When Adam fell, he quickly lost (Newton) 26
When Hannah press'd with grief (Newton) 26
When Israel heard the fiery law (Newton) 26

When Jesus left his Father's throne (Montgomery) 117
When Joseph his brethren beheld (Newton) 5
When like a stranger on our sphere (Montgomery) 69
When our heads are bow'd with woe (Milman)136
Whirlwind for Christ, A (Bayly) 166
Whiston Hospital 172
Whitley Bay Co-operative Society 151
Whitley Bay Peace Week 155
Whitley Seaside Chronicle and Visitors' Gazette, The 152
Whole Psalms, The (Sternhold and Hopkins) 125
Wild Rose: on plucking One late in the Month of October, The (Montgomery) 69, 71
Winchester College 122, 128, *
Wishes that the sluggard frames (Newton) 37, 38
Within Thy courts have millions met (Montgomery) 113
With jubilant united strain (Bayly) 166
With reverence and wonder, we view your work, O God (Bayly) 179
Women Ministers 150
World Alliance for Promoting International Friendship through the Churches 149
World before the Flood, The (Montgomery) 76, 113
World Church Exhibition 160
World Council of Churches 157, 166, 168, 170, 171
Worldling, The (Newton) 33
World Refugee Year 171
World War I 149, 156
World War II 156-162
Worship (Bayly) 174
Wren, The (Montgomery) 111
Yes! since God hinself has said it (Newton) 37
York Castle 66, 67, 68
York Minster 124
Young People's Services 156
Zaccheus (Newton) 33
Zaccheus clim'd the tree (Newton) 33
Zion (Newton) 30

INDEX OF TUNES

This index lists tunes in the text; it does not claim to be a comprehensive list of all the tunes to which particular words have been set. An * indicates the tune is also mentioned in *Hymnwriters* I.

ABBEY 37, 99, 175
ABBOT'S LEIGH 30
ABER 142
ABRIDGE 22
ACH, WANNE KOMMT 83
AJALON 88
ALBANO 142
ALDERSGATE 107
ALLE MENSCHEN MÜSSEN STERBEN 105
ALL SAINTS (German) 31, 37, 56
AMAZING GRACE 6
AMSTERDAM HYMN 166
ANNUE CHRISTE 130, 133, *
ANNUNCIATION 83
AUSTRIA 30, *
AUSTRIAN HYMN 30
BALLERMA 97
BANGOR 99
BELMONT 97, 135, *
BEN RHYDDING 132
BETHLEHEM (Wesley) 83, 95
BISHOP 165
BOHEMIA 107
BOW BRICKHILL 81
BOWDEN 83
BOYCE 37
BRADFIELD 175
BREMEN (Munich) 88
BREMEN (Vulpius) 89
BROCKHAM 81
BROMSGROVE (Dyer) 115
BROOMFIELD 146, 172
BUNESSAN 173
BURFORD 83
CAITHNESS 148, *
CANNONS 131
CANONS 131
CANTERBURY 17, 117
CARING 124
CARLISLE 95
CATON 131
CHRISTUS DER IST MEIN LEBEN 89
CLAPTON 142
COMMUNION (Miller) 131
CONFIDENCE (Clarke) 81
CONTEMPLATION (Ouseley) 175
CRASSELIUS (Hamburg) 94, 112
CRUCIS VICTORIA 115
CRÜGER 88, *
CULBACH 83
DA CHRISTUS GEBOREN WAR 32
DOMINUS REGIS ME 136
DONCASTER 83, 95
DOWN AMPNEY 174
DUKE STREET 107
DUNDEE (Damon) 83
EISENACH 81
ERHALT UNS HERR 131
EVANGEL (Fink) 115
EVELYNS 139
FAIRHOLM ROAD 166
FALCOLN STREET 109
FARRANT 83, *
FENTON COURT 81
FIRST MODE MELODY 83
FLAVIAN 22
FÜR DEIN EMPFANGEN SPEIS UND TRANK 115
GAUDEMUS IGITUR 164
GAUNTLETT 14
GENEVAN PSALM [98] 22
GENEVAN PSALM [118] 22
GETHSEMANE (Monk) 88
GILDAS 83
GILLINGHAM (Clarke) 81
GOD OF GLORY 105
GONFALON ROYAL 94
GOTT DES HIMMELS 37
GOTT SEI DANK 78, 83, 129, *
GRAFTON 81
GRAVETYE 166
HALTON HOLGATE 37
HARTFORD 78
HARTS 78
HERRNHÜT (Crüger) 88
HERRNHÜT (Wesley) 17
HERZLICH THUT MICH VERLANGEN 129, 144
HOLLAND 131
HOLMBRIDGE 88
HOLY CHURCH 88, *
HYMN OF EVE 133
ILLSLEY 165, *
IRIS 81
KENSINGTON NEW 81
KINGSFOLD 117
LA FEILÉE 130, 133
LANCASHIRE (Smart) 88, *
LANGDALE (Redhead) 37, *
LAUDATE DOMINUM (Gauntlett) 14
LAUDS (Wilson) 83
LAUS DEO 167, *
LEEDS 123
LEIPZIG (Schein) 81
LEWES 81
LLYFNANT 88

LONDON NEW 115, 148, *
LOUEZ DIEU 17
LÜBECK 78, 83, 129, *
LUDBOROUGH 107, 113
MACH'S MIT MER GOTT 81
MAINZER 94, 107
MAISEMORE 83
MARNHULL 180
MELCOMBE 107, 165
MINSTER 31
MONKLAND 132
MONTGOMERY (Goodall) 95
MONTGOMERY (Woodbury) 112
MUNICH 88
NATIVITY (Lahee) 95
NEW BRITAIN 6
NEWBURY (Chatham) 94
NEWINGTON (Jones) 22
NIAGARA 81
NICHT SO TRAURIG (Bach) 88
NICHT SO TRAURIG (Freylinghausen) 88, *
NORTHAMPTON 83
NORTH PETHERTON 174
NORTHUMBRIA 164
NORWOOD (SA) 88
NOX PRAECESSIT 83
O HAUPT VOLL BLUT UND WUNDER 129, 144
OLD 25TH 112
OLD 50TH 112
OLD 100TH 99
OLD 104TH 14
OLD 132ND 22
OLD 134TH 95, 123
ORIENTIS PARTIBUS 32
OSWALD'S TREE 22
PALANTINE 83
PALMS OF GLORY 104
PASSION CHORALE 129, 144
PASTOR 89
PETRA 88
PHILADELPHIA 131
PRAETORIUS 115
PRESSBURG 88, *
PSALM [136] 17
QUAM DILECTA 130
RACHEL 31
RAVENSCROFT 14
RAVENSHAW 132
REDHEAD NO. [46]167
REDHEAD NO. [76] 88, *
REGENT SQUARE 83
REJOICE O PEOPLE 164
RENDEZ Á DIEU 22, *
RESURREXIT 142
RHODES 109
RICHMOND (Stephens) 17
RILEY 83
ROCHESTER 83
ROCKINGHAM 131
RUSSIA (Bortnianski) 94, 112
SAINT AMBROSE (Dykes) 37
SAINT ANTHOLIN 167
SAINT AUGUSTINE 83
SAINT BARTHOLOMEW (Duncalf) 94
SAINT BERNARD (Tochter Sion, adapted Richardson) 31, *
SAINT BOTOLPH 31
SAINT CHRYSOSTOM 37, *
SAINT DAVID (Ravenscroft) 115
SAINT ETHELWALD 83
SAINT FLAVIAN 22, *
SAINT FRANCES 83
SAINT GEORGE (Elvey) 83
SAINT GEORGE (Gauntlett) 132
SAINT GEORGE'S WINDSOR 83
SAINT GREGORY 131
SAINT HELENA 124
SAINT HUGH (Hopkins) 83, 175
SAINT JOHN BAPTIST (Calkin) 175
SAINT LEONARD (Bach) 31, 56
SAINT MARTIN (Old French Melody) 32
SAINT MATTHEW 178
SAINT MATTHIAS (Gibbons) 83
SAINT MICHAEL 95, 123, *
SAINT OLAVE 132
SAINT OSWALD 37
SAINT PETER 31, 83
SAINT PETERSBURG 94, 112, *
SAINT POLYCARP 78
SAINT STEPHEN (Jones) 22
SAINT STEPHEN (Smith) 22
SAINT TIMOTHY 142
SALZBURG (Hintze) 105
SANDYS PSALM [8] 78
SAN ROCCO 147, 148
SAVANNAH (Herrnhüt) 17, *
SCHEIN 81
SCHÖNBERG 105
SHARON (Boyce) 37
SHARPTHORNE 166
SHEEN 173
SHEPHERDS IN THE FIELDS 81
SILVER STREET 109
SIMPLICITY (Gibbons) 17, 117
SLEEPER, AWAKE 106
SONG [1] 164
SONG [13] 17
SONG [34] 107
SONG [67] 83
SOUTHWELL (Irons) 175
SPAIN, THE 179
SPETISBURY 14
STRACARTHO 31, *
STUTTGART 167, *
SWABIA 107, 132

TALLIS 97, 99
TALLIS NINTH TUNE 97, 99
TALLIS' ORDINAL 97, 99
TANTUM ERGO (French Melody) 81
TANTUM ERGO SACRAMENTUM 81
THANKSGIVING (GILBERT) 83
THAXTED 176, 179
THEODORA (Handel) 17
THE SPAIN 179
TICHFIELD 105
TITCHFIELD 105
TIVERTON 95
TOURS 144
TUNBRIDGE 17
TYES CROSS 166
UNIVERSITY COLLEGE 130
UXBRIDGE 133, *
VENICE 83
VERMONT 165
VIENNA (Pleyel) 130
VULPIUS (7676) 89
WACHET AUF! 106
WAINWRIGHT 88, 107
WALTHAM 37
WARRINGTON 107, 165, *
WATCHMAN (Leach) 83
WELLS (Bortnianski) 94, 112
WELLSPRING 94, 112
WER NUR DEN LIEBEN GOTT 94, 112
WESTFIELDS 165
WETHERBY 22
WHITEHALL 78
WIGTON 83
WIGTOWN 83
WINCHESTER NEW 94, 112, *
WINDSOR 83, *
WOODFORD GREEN 81